室内细部图集

商店与住宅

INTERIOR DETAIL SHOP & RESIDENCE

凤凰空间　编

细部节点 **100**例

江西科学技术出版社

Contents

餐厅

商店与住宅

办公场所与教育机构

咖啡馆

医疗中心与文化中心

酒店与休闲场所

*本书中所有无单位数值均为毫米。

商店
辣潮　6
荷玛家　16
斯佩罗　24
利莫克　34
伊蒂之屋　42
考比勒　52
骊住展厅　60
"试衣间"　70
迪普美发沙龙　80
金宝热轨书屋　90
里科塔奶酪沙拉　104
安利广场　110

住宅
永恒　130
帕克力拓住宅　140
H住宅　150
顶层公寓住宅　160
歌劳斯　168
马山住宅　176

商店

辣潮

荷玛家

斯佩罗

利莫克

伊蒂之屋

考比勒

骊住展厅

"试衣间"

迪普美发沙龙

金宝热轨书屋

里科塔奶酪沙拉

安利广场

| 辣潮 | 建筑面积：177平方米

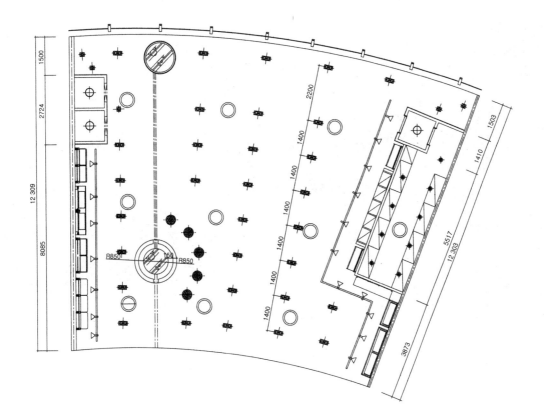

天花图

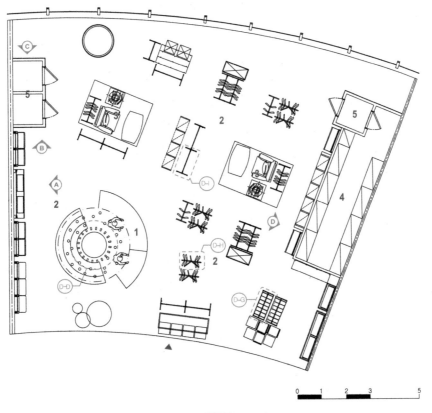

1. 收银台
2. 展示区
3. 座位区
4. 仓库
5. 试衣间

平面图

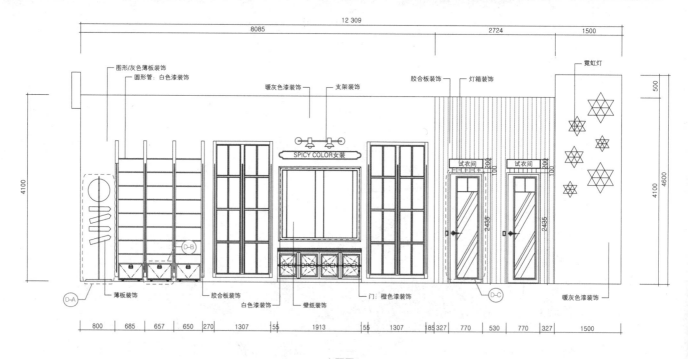

立面图A

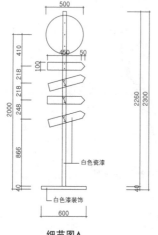

细节图A

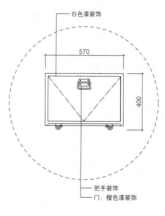

细节B

■ 试衣间

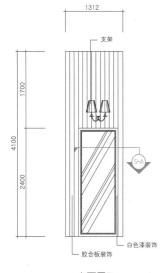
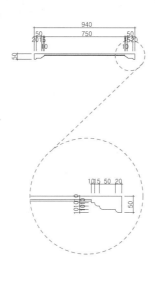
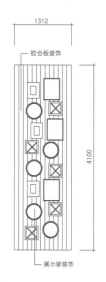

立面图B 剖面图A 立面图C

■ 细节图C——试衣间门

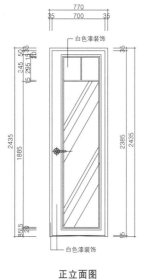
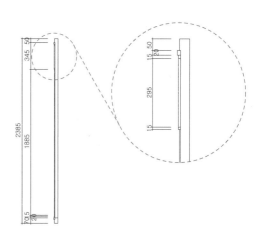

正立面图 剖面图

■ 细节图D——展示柱

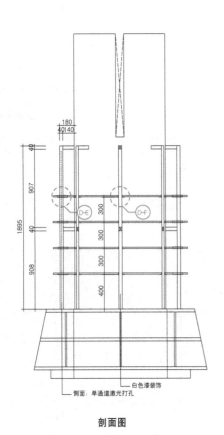

剖面图

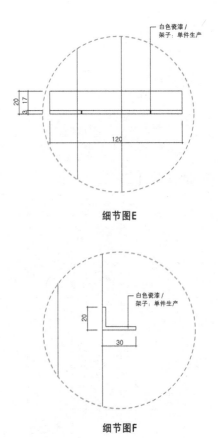

细节图E

细节图F

PICY COLOR, uses a wide variety of colors to create an eye-catching ght and pleasant atmosphere under the slogan of the brand 'Fashion is un Play & Pleasant Imagination'.

he entire space creates a shopping environment using 4 colors such as range, blue, white and wood for free and pleasant shopping. Added with rniture of different designs, cute graphics and stripe patterns on floor d wall, cozier and richer environment is created in this space. In partic- ar, unlike common fashion shop, the cashier counter is arranged in the iddle interconnected with the column to create it as an object, around ich products are arranged freely so that the traffic lines of customers e organically constructed.

his completes a pleasant shopping space with fun elements which high- ghts brand identity.

辣潮，在内部装修上运用了多种的颜色搭配，炫目的灯光营造出一种愉悦的氛围。该品牌的宣传语为"时尚是有趣的玩乐及愉悦的想象"。

通过使用橙色、蓝色、白色及原木色四种颜色，为整体空间营造出一种愉悦且自由的购物环境。再加上设计风格各异的家具、可爱的图形、经色条修饰的墙体及地板，整个空间充满了温暖的感觉。与其他常见的时装店不同的是，辣潮将收银台设置在与各个商品柜相连的中心，其周围所有的产品陈列位置自由随性，将客人的移动路线有效地组织起来。

如此有趣的理念及设置，更好地完善了购物环境的愉悦性，从而凸显了品牌的形象。

■ 细节图G——衣架1

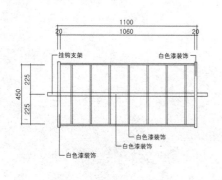
俯视图

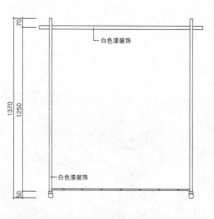
正立面图

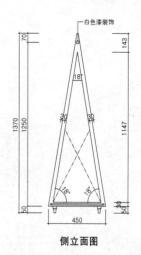
侧立面图

■ 细节图H——衣架2

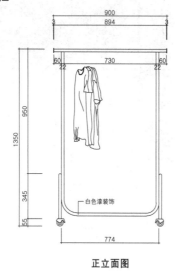

正立面图　　　　侧立面图

■ 细节图I——衣架3

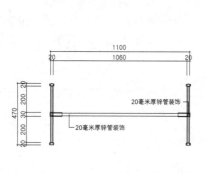
俯视图

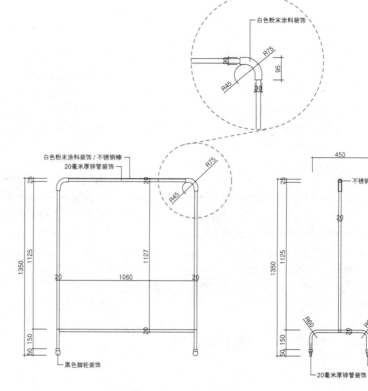
正立面图　　　　剖面图

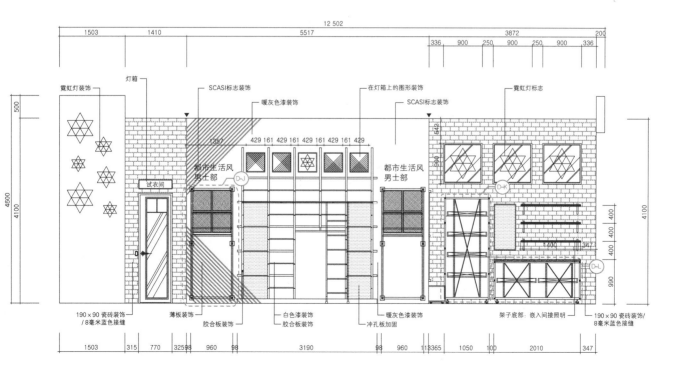

立面图D

■ 细节图J——框架板

俯视图

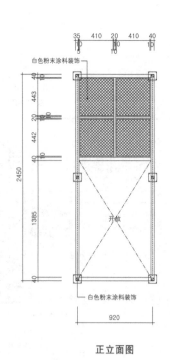

正立面图

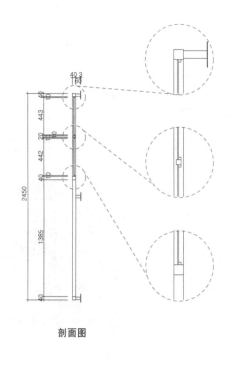

剖面图

细节图K——墙面支架1

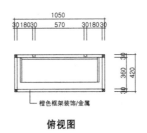

俯视图

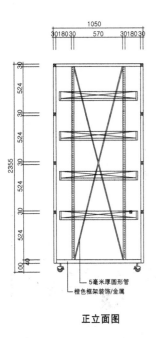

正立面图

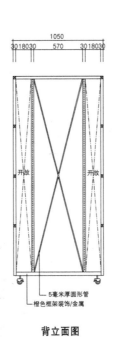

背立面图

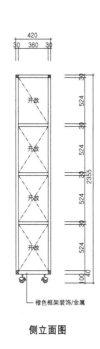

侧立面图

细节图L——墙面支架2

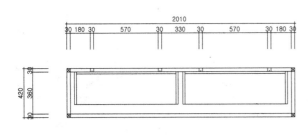

俯视图

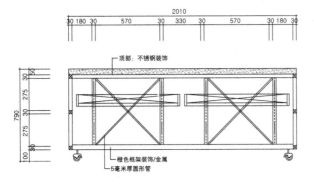

正立面图

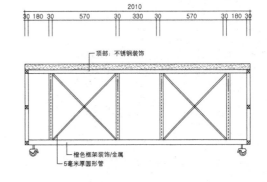

背立面图

荷玛家

建筑面积：162平方米

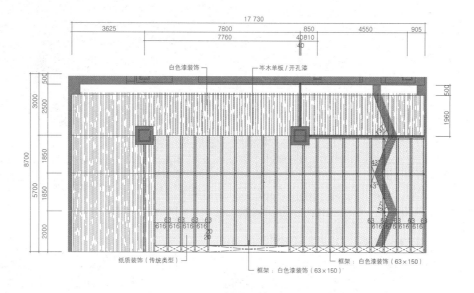

屋顶平面图

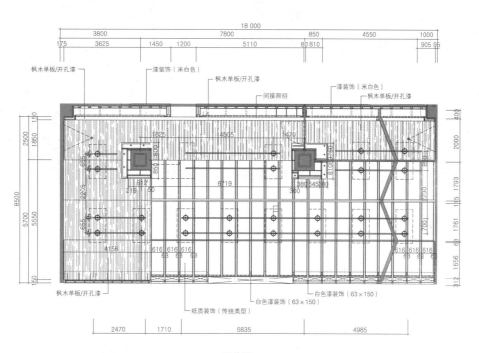

天花图

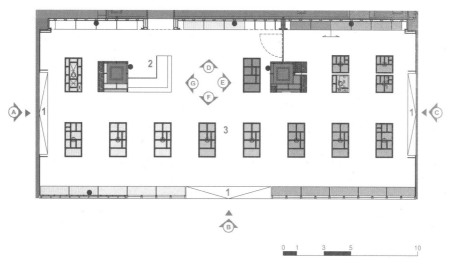

1. 入口
2. 收银台
3. 展厅

平面图

Heim, which means 'to home' in German. The lifestyle shop selling natural and healthy products provides a comfortable space for people to take a break from their hectic daily life. Inspired by leading Portuguese architect brothers Aires Mateus, the designer reduced ornamental elements, and focused on vertical and horizontal structure.

Under the concept of a 'house inside a house', the gabled roof structure gives a feeling as if entering from a space into another. The exterior wall finished with wood frame and glass divides the space, and at the same time opens up the interior to diminish the stifled feel of a small space. The use of wood and white coating in the sales and display area in the back creates a neat and balanced ambience, while the pointed roof presents a cozy feel of an attic. The display tables appear to be floating in the air due to the built-in indirect lighting on the floor. The entrances are on three sides to increase the accessibility, and the natural beauty emphasizes the quality of products and elevates the values of the brand.

Heim，在德文中是"到家"的意思。这间注重生活方式的店铺主出售纯天然及健康产品，让人们从奔波劳累的日常生活中解放出来为他们提供一个舒适的购物环境。受葡萄牙先驱设计师艾利斯·特乌斯兄弟的启发，设计师减少了室内装饰性元素，更加专注于直与水平结构。

为了实现"房中房"的理念，设计师采用人字形屋顶结构使人产一种从一个空间进入另一个空间的感觉。外景墙则通过运用木质架及玻璃的修饰来划分区域，与此同时，使室内空间更显开放，减少小空间带来的窒息感。商品及展示区背面的木质产品及白色层的使用，创造出一种整洁、平衡的氛围。同时，尖角屋顶也给一种阁楼般安逸的感觉。由于地面上嵌入式间接照明的效果，整展示台看起来犹如飘浮在空中一般。整个店铺三面都设置了入口为顾客出入提供了更多的便利。纯天然制品的魅力强化了产品的量，提升了该品牌的价值。

■ 入口

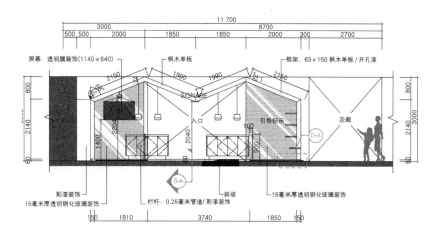

立面图A

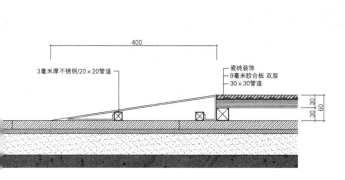

剖面图A

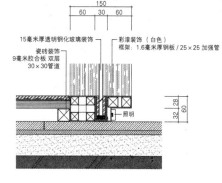

细节图A

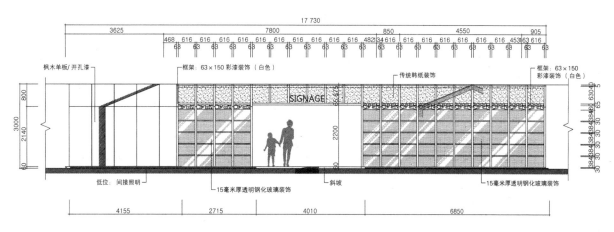

立面图B

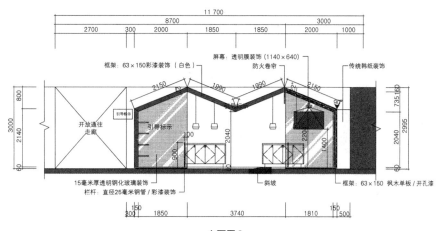

立面图C

■ 展厅

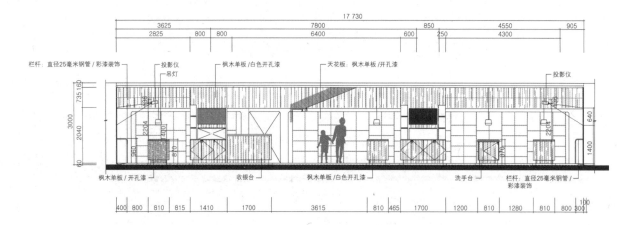

立面图D

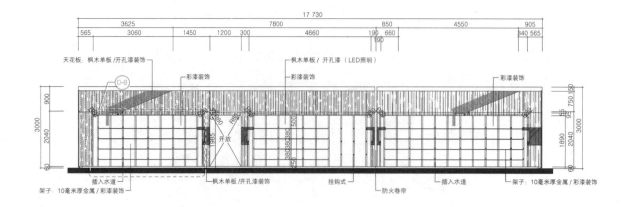

立面图D'

细节图B——墙架

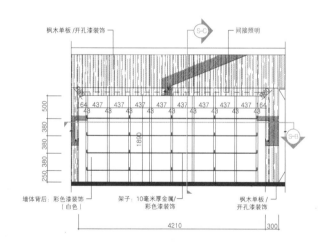

正立面图

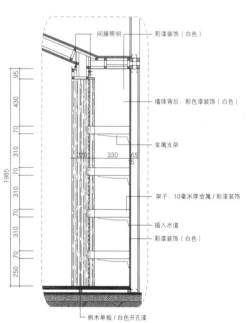

剖面图C

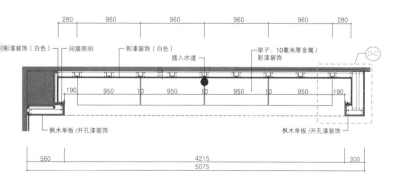

剖面图B

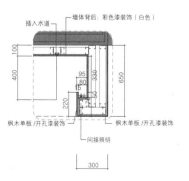

细节图C

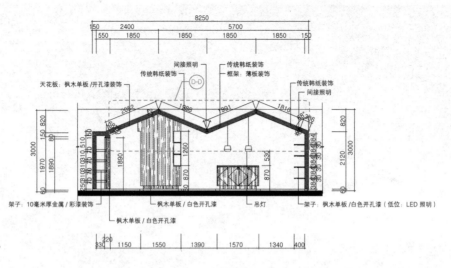

立面图E

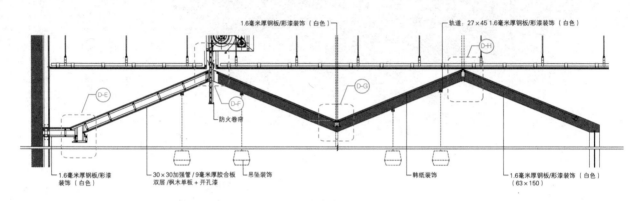

细节图D

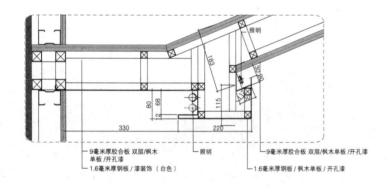

细节图E 细节图F

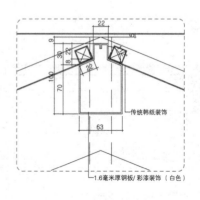

细节图G 细节图H

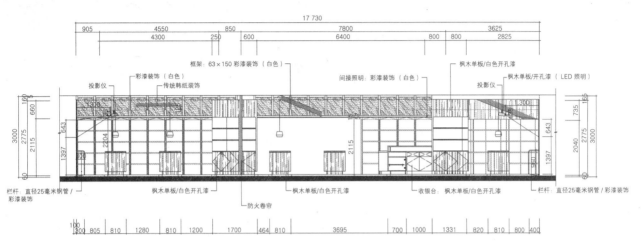

立面图F

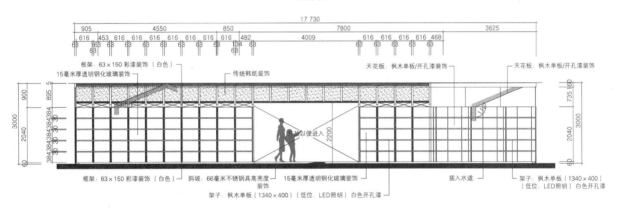

立面图F'

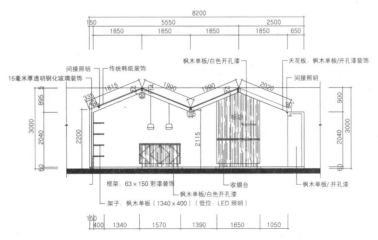

立面图G

斯佩罗

建筑面积：61平方米

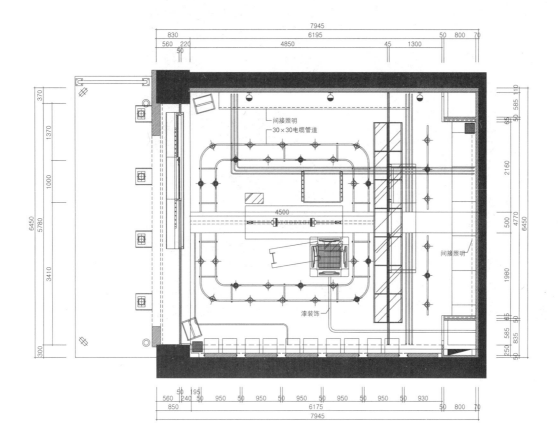

天花图

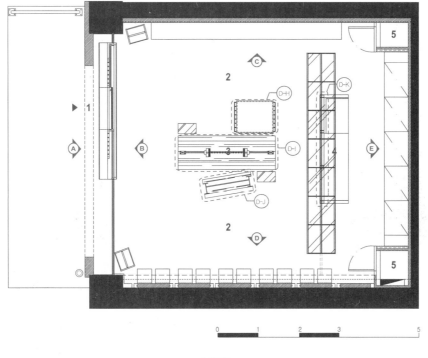

1. 入口
2. 礼堂
3. 展台
4. 收银台
5. 仓库

平面图

■ 外观

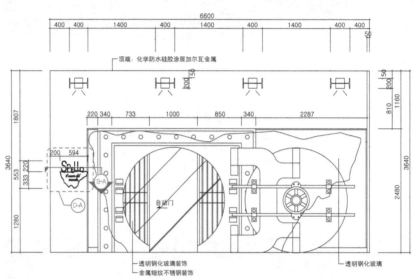

立面图A

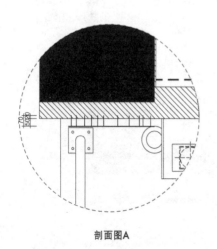

剖面图A

细节图A

入口

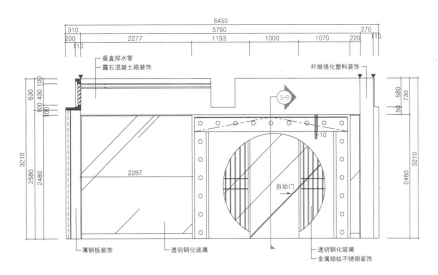

立面图B

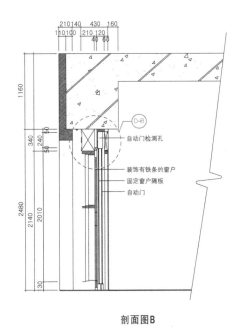

剖面图B

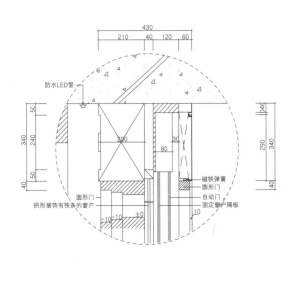

细节图B

■ 展示区

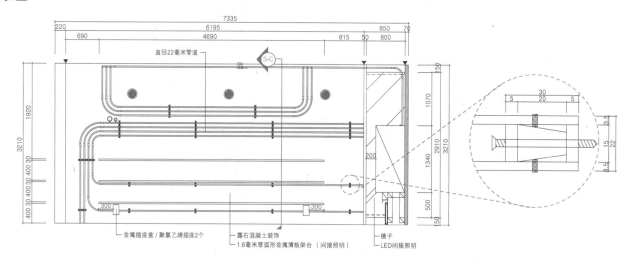

立面图C

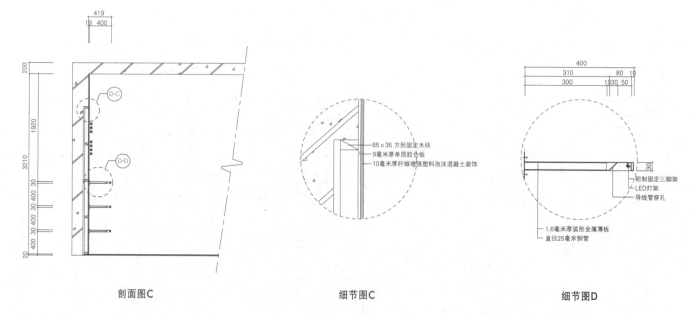

剖面图C　　　　　　细节图C　　　　　　细节图D

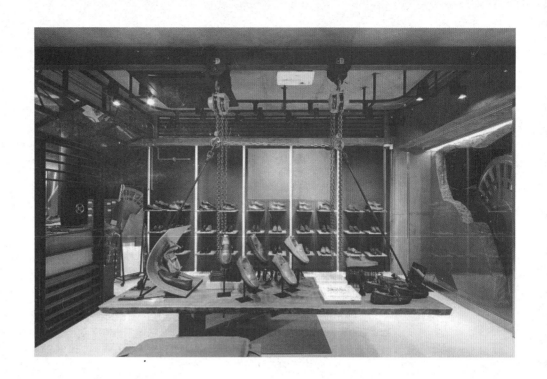

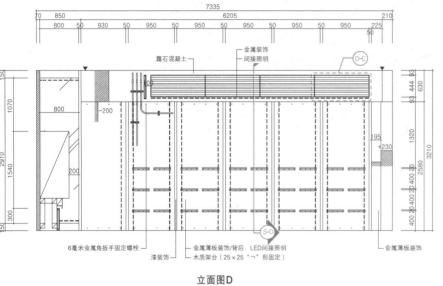

立面图D

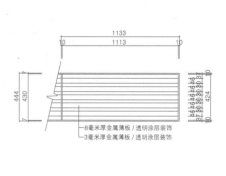

细节图E

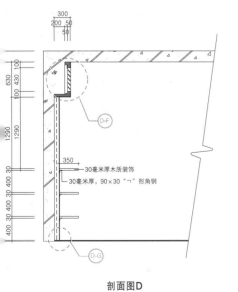

剖面图D

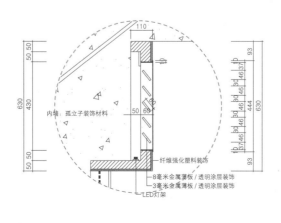

细节图F

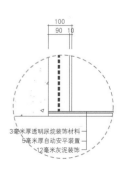

细节图G

SPELLO, men's shoes brand.

Under the concept of 'Vault of safe', the unique exterior catches the attention using a huge safe usually found in a bank. On the outside, the safe door is installed in the broken down reinforced concrete to reflect the brand's philosophy that treasure the shoes as precious metals and valuables, inducing curiosity and tension. The inside of the safe has been designed in an industrial style with exposed concrete, steel, and pipes. On the center is an island display shelf made of construction equipment for a rough and dynamic feel. In the back of the counter reinforced with steel structure is a VIP locker that looks as a safe deposit, emphasizing the uniqueness and presenting a contrast against the hall.

The sensuous design that differentiates itself from existing shoes stores offers a special experience to the customers.

斯佩罗，是一个男鞋品牌。

设计师受到"保险库"的启发，为了吸引大家的注意使用了银行里大保险库的独特外形。从商店外面看，"保险库"的门安装在损毁钢筋混凝土墙中。这样特别的设计体现了该品牌"请把鞋子当成贵重物品来珍藏"的理念，并增加了品牌趣味性和张力。商店的内部被设计成充斥着混凝土、钢筋及管道的工业风格。在整个商店的中部是一个用建筑设备建造的中岛形的展示架，给人一种原始动态的感觉。固金属结构的柜台背后是 VIP 带锁储物柜，看起来像是一个保险箱，这样的设计既强调了独特性，又体现了与展厅的反差。

这种极具美感的设计使该品牌从常见的鞋店脱颖而出，给顾客一种别的购物体验。

细节图H——凳子

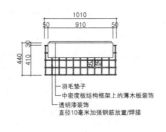
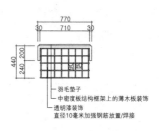

俯视图　　　　　　　　　　正立面图　　　　　　　　　　侧立面图

细节图I——展台

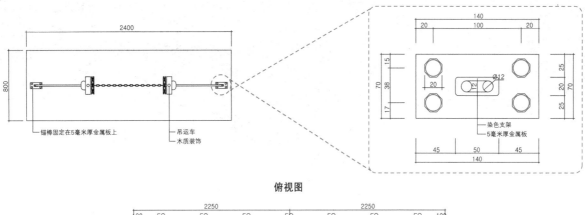

俯视图

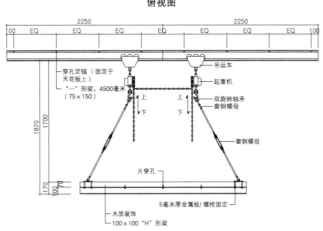

正立面图

细节图J——木头凳子

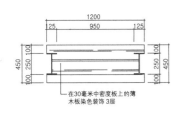
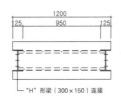
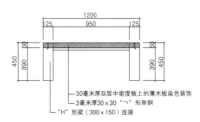

俯视图　　　　　　　　　剖面图（俯视图）　　　　　　　正立面图

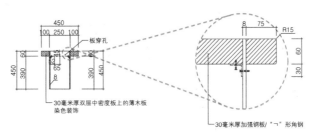

侧立面图

■ 细节图K——柜台框架

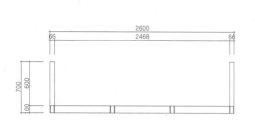

俯视图

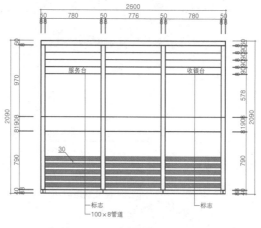

正立面图

侧立面图

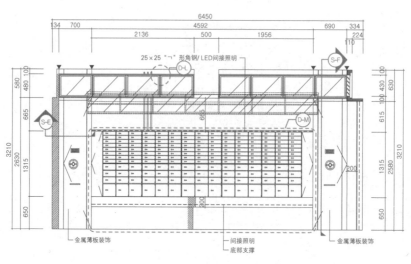

立面图E

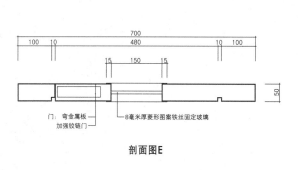

剖面图E

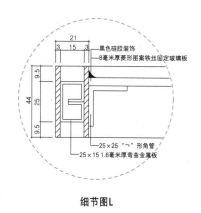

细节图L

细节图M——保险库

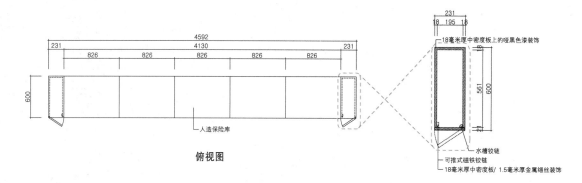

俯视图

正立面图

剖面图F——仓库

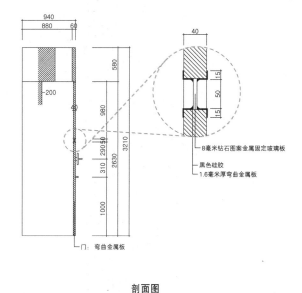

剖面图　　　　左剖立面图　　　　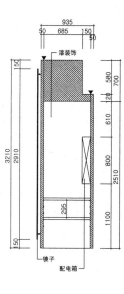　右剖立面图

利莫克

建筑面积：30平方米

■ 表面

■ 细节图A——展台

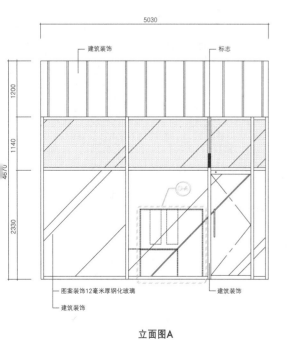

立面图A

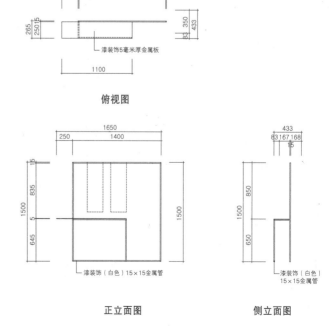

俯视图

正立面图　　侧立面图

1. 入口
2. 展示区
3. 收银台
4. 仓库

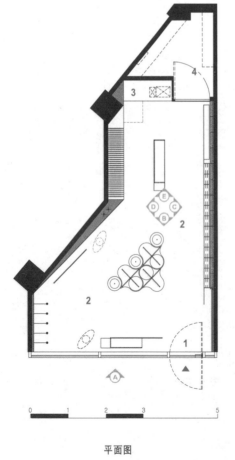

平面图

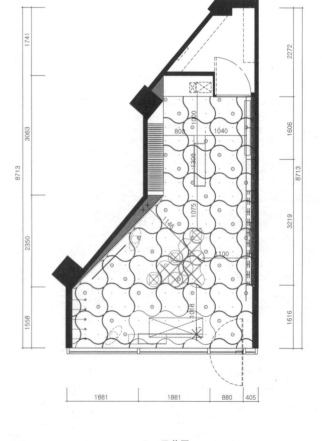

天花图

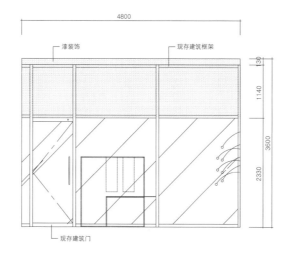

立面图B

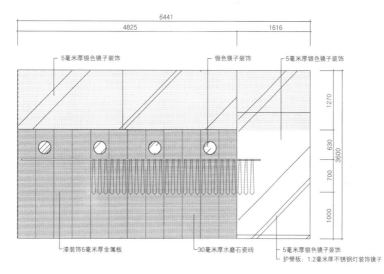

立面图C

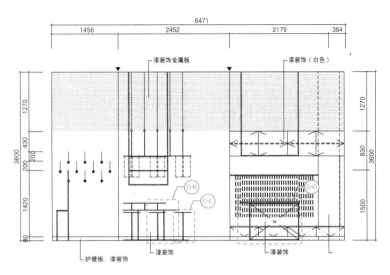

立面图D

细节图B

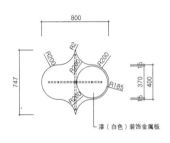
俯视图

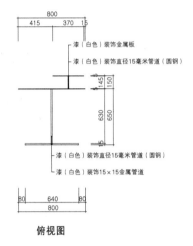
俯视图

侧立面图

细节图C

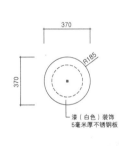
俯视图

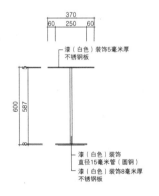
正立面图

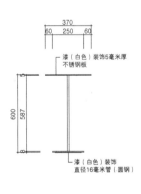
侧立面图

细节图D

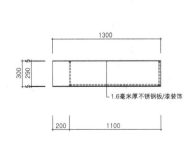
俯视图

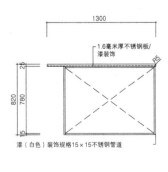
正立面图

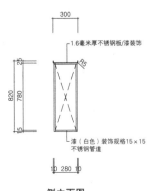
侧立面图

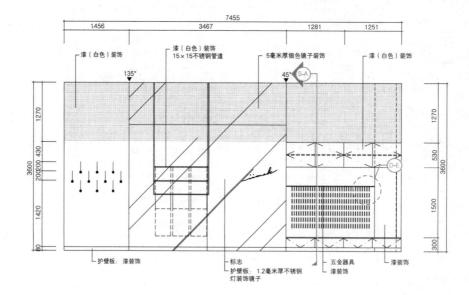

立面图D'

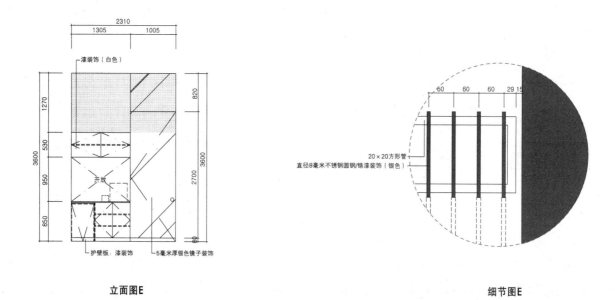

立面图E

细节图E

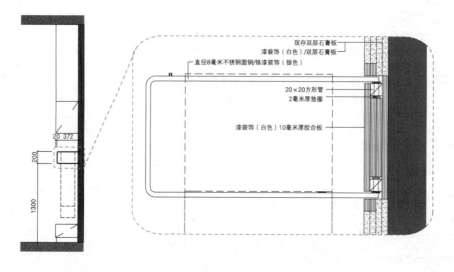

剖面图A

Limok was inspired by Chopsticks Gallery, the latest work by the designer, and wanted to create a space providing healing and relaxing. The designer supervised the project from the brand design and naming to details, conducting a thorough transformation of the space into a whole new one.

Standing for 'drawing the attention', the name 'limok' written on the signage look as if the characters are woven, representing the atmosphere of the space. The keywords for the interior are soft, light, flexible, organic, weaving, formless, and bright. The inside is finished by undulating fabric of the ceiling, and the achromatic materials that highlights the colors of the products on display. The cloud-like display tables and hangers in unique forms add a sense of amusement to the rather small space, which is visually expanded by the mirror on the wall. The mirror is installed in an angle of 45 degree from the slanted wall, creating an illusion of a rectangular space.

The designer completes the space with a unique and peaceful atmosphere, reflecting the demand of the client who wishes to emphasize the space itself.

在筷子展览馆最新设计的启发下，利莫克想要营造出一种治愈休闲的空间。设计师从品牌设计、品牌名称等各个方面严格把关，将整个商店转换成一个新的空间。

作为"最具吸引力"的代表，"Limok"这个印在标志上的名字，每个字母看起来像是针织上去的，以了更好地体现该商店特殊的空间氛围。内部装饰以柔软、轻快、有机、编织、无形且明亮为特点。主要特点是在天花板上装有呈波浪形的布料，在展区用无色的材料与产品的多彩形成鲜明对比。云朵般形状的特殊展柜及衣架，为这个相对较小的空间增加了趣味性。而墙上的镜子，在视觉上使店面看起来有所增大。镜子以45度角倾斜的方式安装在墙上，创造出一种矩形的空间幻觉。

设计师最后以一种独特而平和的氛围完成了的设计，以满足消费者们对商店空间本身的期待。

伊蒂之屋

建筑面积：108平方米

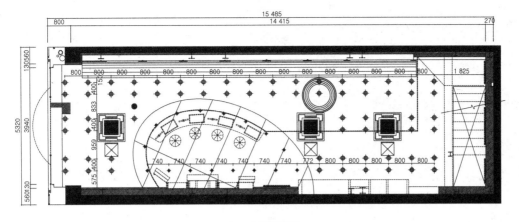

二层天花图

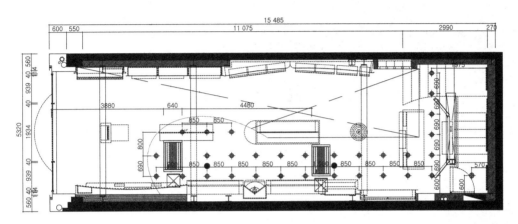

一层天花图

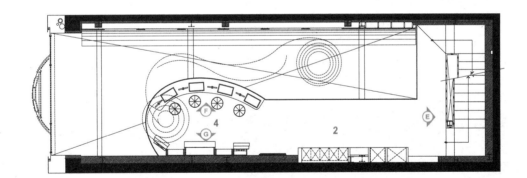

1. 入口
2. 展示区
3. 收银台
4. 平板电脑区

二层平面图

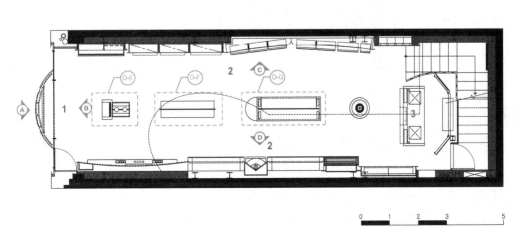

一层平面图

伊蒂之屋

Under the concept of 'a princess in reality', the first branch ETUDE HOUSE was designed as a cheerful experiential cosmeti store.

The four-level-high facade recreates a castle in a tale, and th pink ribbons connect the exterior to the interior, guiding the cu tomers inside in a natural manner. The girl-like interior enriched by charming origami-like shelves, fixtures, and ligh ings. The staircase adorned with mirrors and paper art objec leads to the experience area upstairs such as a crown zone and tablet pc zone, where the shelves, fixtures, and stainless ste mirror on the ceiling break down the boundary between spaces.

The functionality and the unique spatial design that reflects th sensibility of the customers represent the identity of the brand.

伊蒂之屋的设计理念为"现实中的公主",第一家店被设计成一个让人容易产生愉悦感的美妆店。

店面总共四层,外观装饰成童话中城堡的样子,并用粉色带将内部与外部联系起来,使顾客自然而然地走入店中。内部设计大量使用了折纸风格的架子、家具及灯光,使得商内部装饰更受女孩子们的喜爱。通过以镜子及纸质艺术品饰的楼梯间就可以到达楼上的体验区、皇冠区和电脑区。子、家具及天花板上的不锈钢镜子使各个区域无缝连接。

这种功能性与独特的空间设计相结合的设计更好地反映出客对该品牌定位的敏感性。

正面

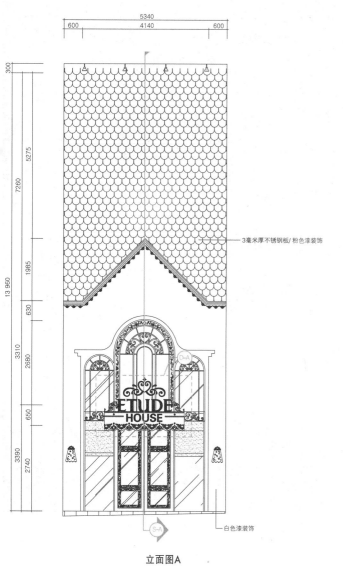

立面图A 剖面图A

■ 细节图A——标志

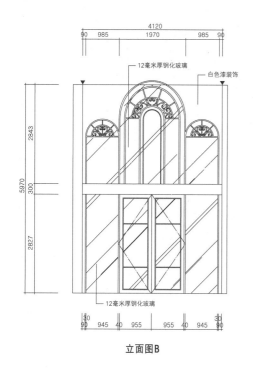

立面图B

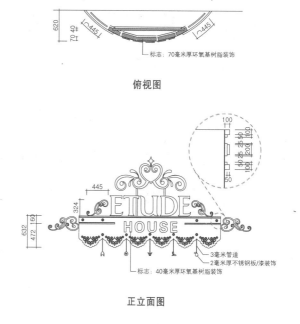

俯视图

正立面图

伊蒂之屋

■ 展示区

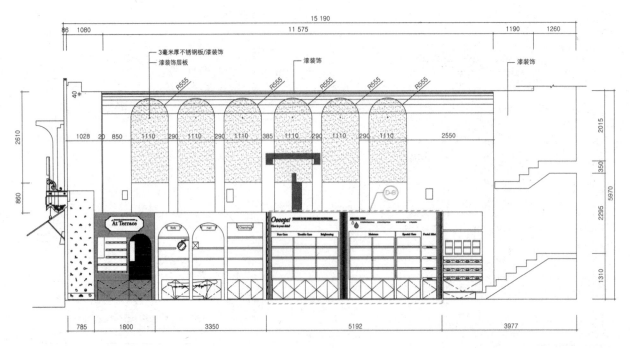

立面图C

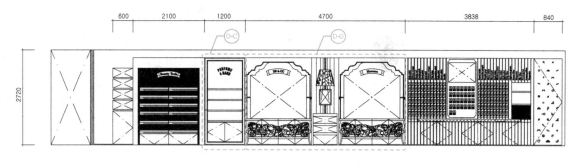

立面图D

细节图B

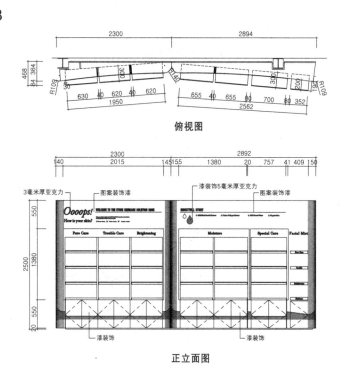

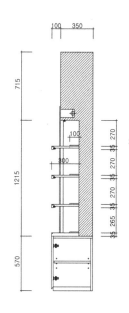

细节图C

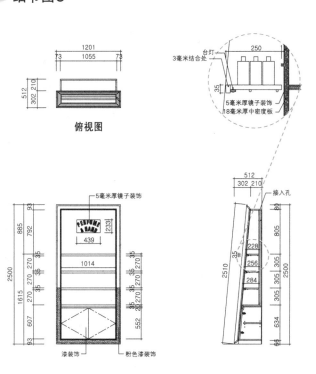

细节图D

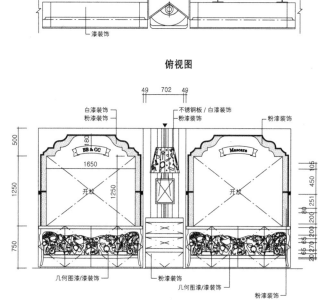

■ 细节图E——促销台

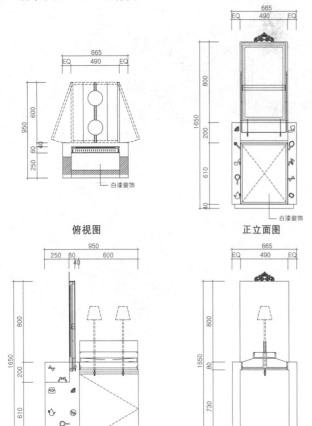

俯视图　正立面图　侧立面图　背立面图

■ 细节图F——口红展台

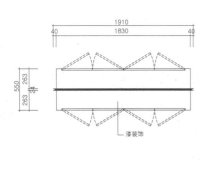
俯视图

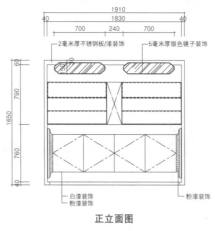
正立面图

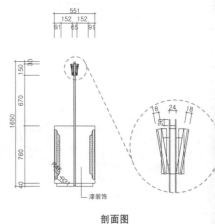
剖面图

■ 细节图G——眼部/面部用品台

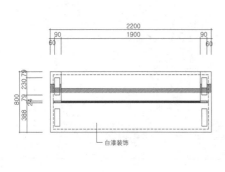
俯视图

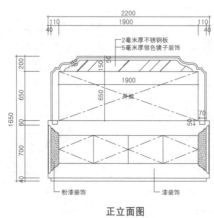
正立面图

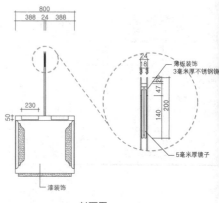
剖面图

楼梯间

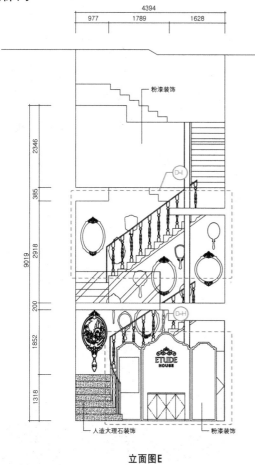

立面图E

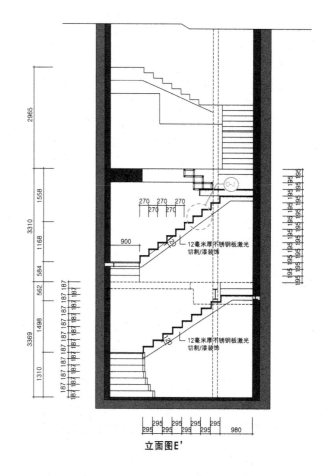

立面图E'

细节图H

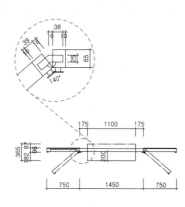

俯视图

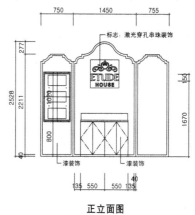

正立面图

细节图I

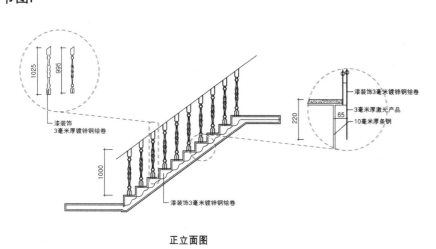

正立面图

细节图J

伊蒂之屋

■ 平板电脑区

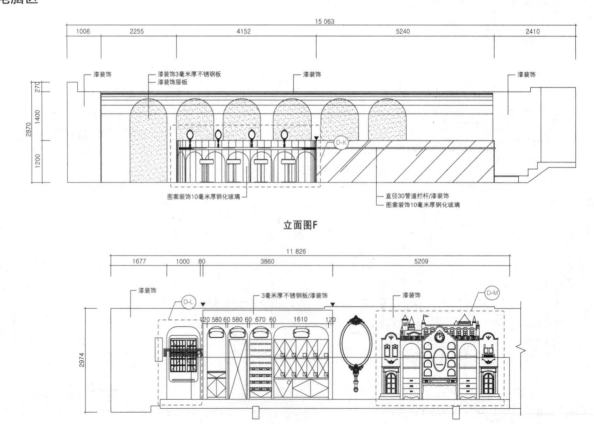

立面图F

立面图G

细节图K——电脑桌

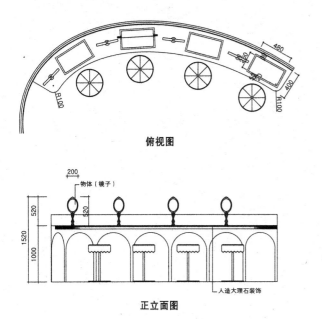

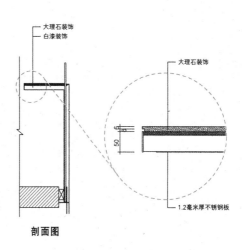

细节图L

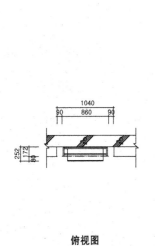

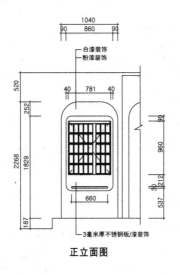

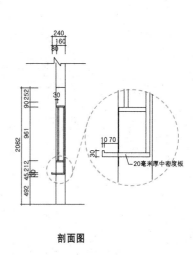

细节图M

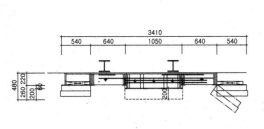

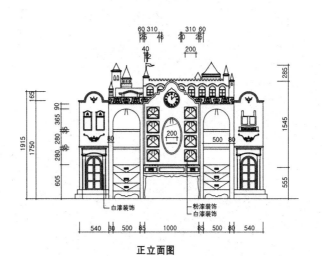

考比勒

建筑面积：68平方米

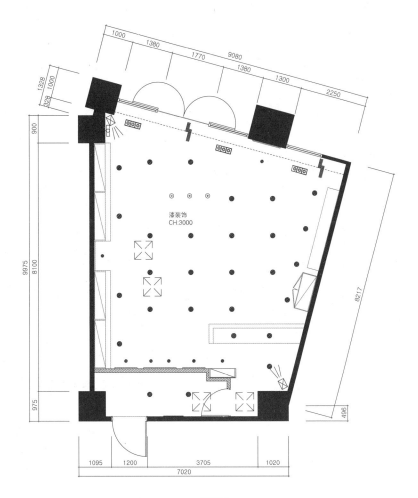

天花图

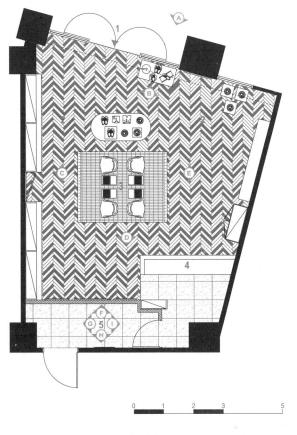

1. 入口
2. 展示区
3. 试衣间
4. 收银台
5. 储藏室

平面图

■ 入口

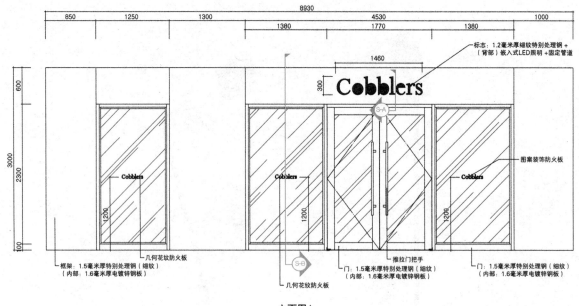

立面图A

剖面图A

剖面图B

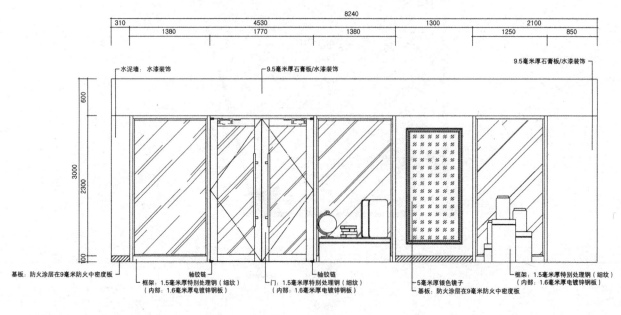

立面图B

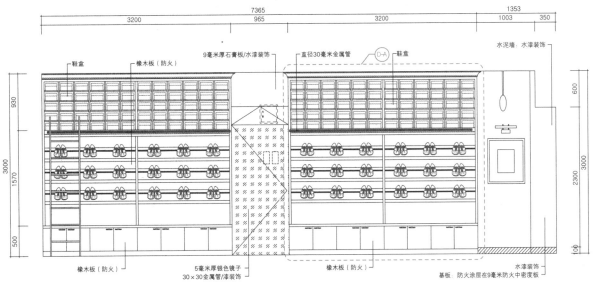

立面图C

细节图A——鞋柜

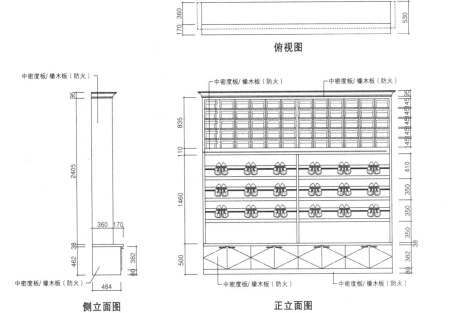

侧立面图　　　　正立面图　　　　剖面图

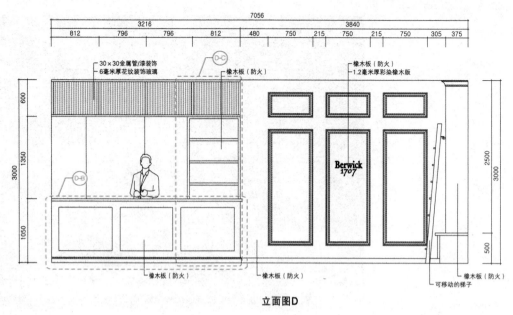

立面图D

■ 细节图B——柜台

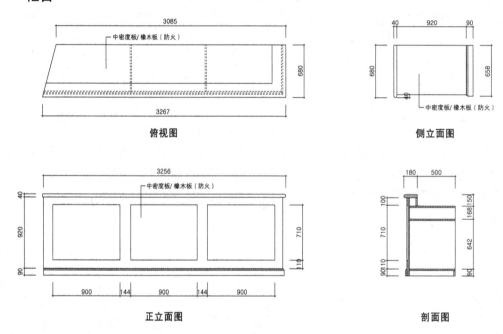

俯视图　　　　　　　　　　　　侧立面图

正立面图　　　　　　　　　　　剖面图

■ 细节图C——墙上的家具装饰

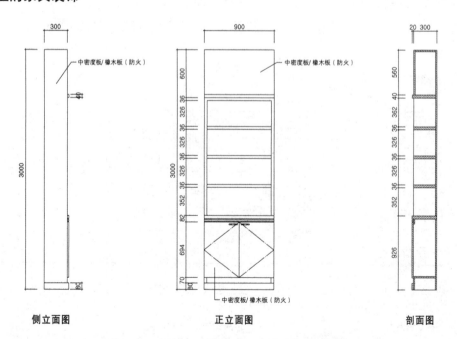

侧立面图　　　　正立面图　　　　剖面图

obblers provides a relaxing space in renewed COEX crowded with people. Since it rgets at those who are familiar with Berwick rather than the general public, the shop as designed to effectively deliver the identity of the brand.

he neat signage leads to the inside which is finished with wood to present a classical mosphere. The dark wood materials and elegant design give a sense of weight and uality. A floor-to-ceiling cabinet for display and storage covers a wall, while display bles and shelves of various sizes achieve an efficient sales space usage.

考比勒为重建后人潮拥挤的会展中心提供了一个放松的空间。店铺针对那些熟悉 Berwick 品牌的老客户，而不是普通大众。商店的设计更有效地传达了品牌的特性。

整洁的标志将客人引入店中，整个店铺通过木质的装饰给人一种经典的感觉。深色木质材料的使用及优雅的设计给客人一种沉重的质感。从地面到天花板的展品陈列柜及储藏室布满墙面。同时，大小各异的展台及架子更充分地利用了店铺的空间。

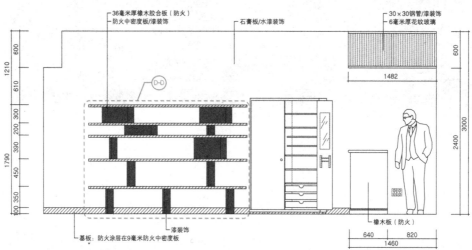

立面图E

■ 细节图D——墙上的家具装饰

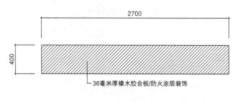

俯视图

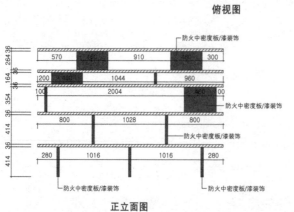

正立面图　　　　　　侧立面图

储藏室

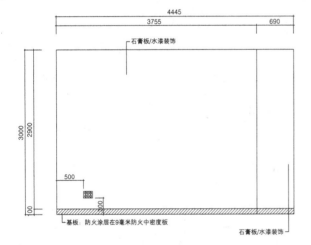

立面图F

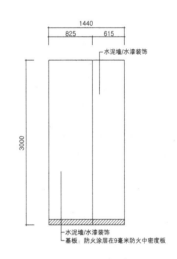

立面图G

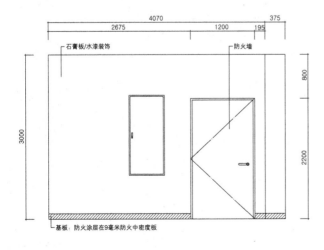

立面图H

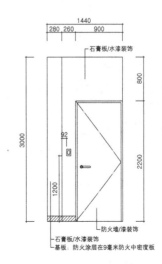

立面图I

| 骊住展厅 | 建筑面积：335平方米

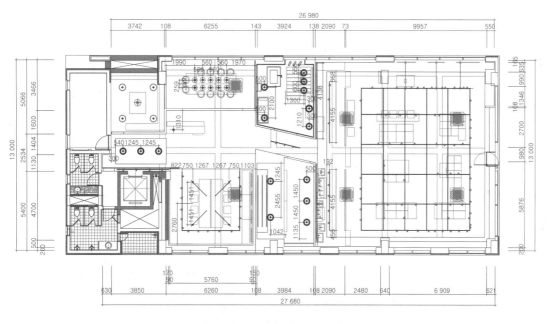

天花图

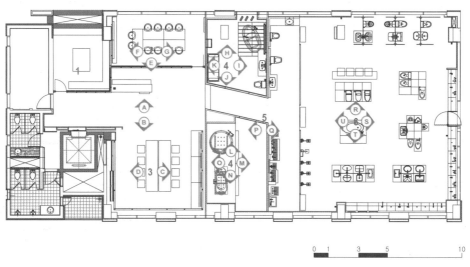

1. 样品间
2. 会议室
3. 骊住展区
4. 亚科斯森展区
5. 特享专区
6. 美标展区

平面图

■ 骊住展区

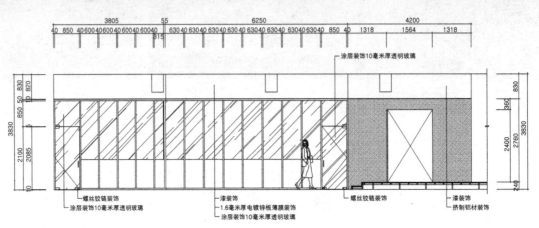

立面图A

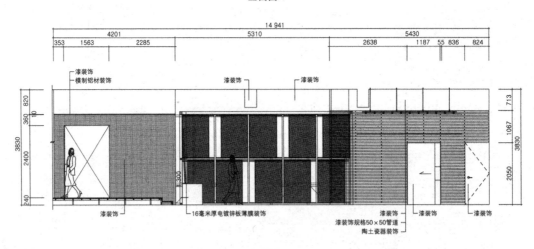

立面图B

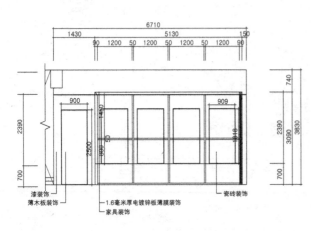

立面图C

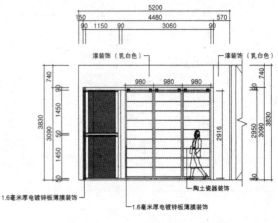

立面图D

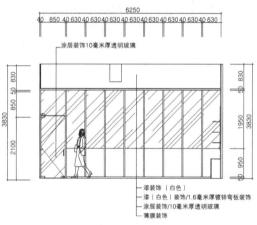

立面图E

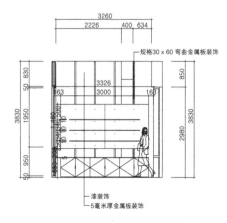

立面图F

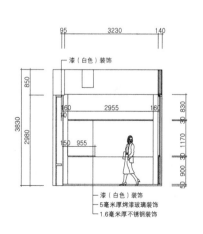

立面图G

骊住展厅

■ 亚科斯森展区

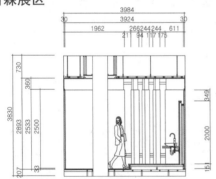

立面图H

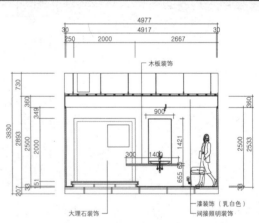

立面图I

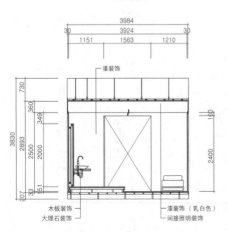

立面图J

立面图K

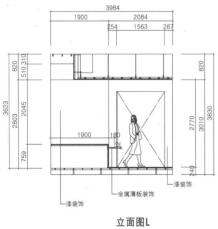

立面图L

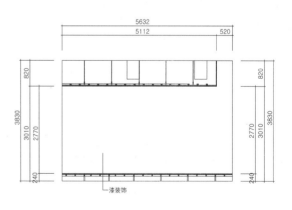

立面图M

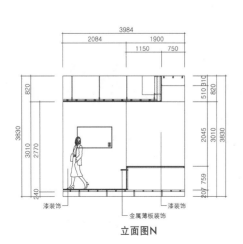

立面图N

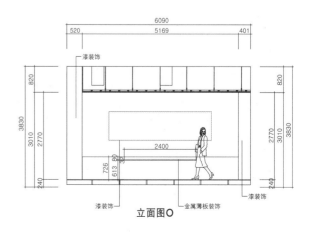

立面图O

■ 特享专区

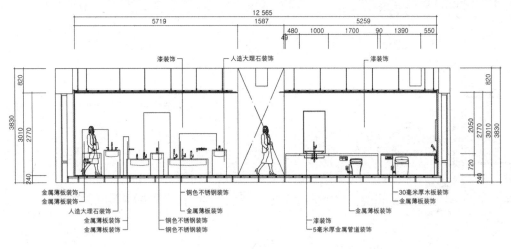

立面图P

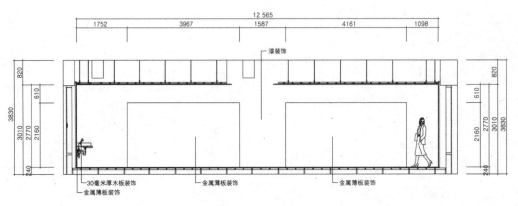

立面图Q

e new showroom for Lixil Group features its representative brands including
nerican Standard, Inax, and Jaxson. The showroom consists of five brand zones
h unique design and materials.

e Lixil zone with interior and exterior tiles themed with the Earth opens up the
rney to the inside. The customers can examine the texture, color, and pattern of
rchandise hung on the walls, as well as the samples on the large central table
ore applying them to their own spaces.

yond the Lixil zone is a Jaxon zone separated into two zones. The standing type
throom reminds of the one of a luxury resort, while the lighting adds a depth to
other bathroom with a premium whirl pool.

ght behind the Jaxon zone, Inax zone displays products including an integrated
et, high quality toilet, and a washstand. The partitions made of hot rolled steel
sets a closed feeling, and the stone and metal fixtures for product display gives a
nse of luxury.

e American Standard zone, which takes up the large area in the furthest back, is
ided into a bathroom collection and a display area. The representative basin mix-
 and ceramic wares are displayed on separate fixtures according to various cate-
ries. The collection zone on a wall shows quality bathrooms of five different con-
ots, and is separated by thin metal frames.

骊住集团的展厅展示了其代表品牌，包括：美标、伊奈和亚科斯森。展厅由5个品牌区域构成，均运用了独特的设计和材料。

骊住展区的内部和外部的瓷砖设计都以"通向地球内部的旅程"为主题。顾客们在购买产品之前，可以检验挂在墙上和陈列在中央大展台上的产品的质地、颜色及图案。

通过骊住展区，可以到达亚科斯森的一个展区，该品牌展区被分成两个区域。标准样式的浴室让人想起了豪华度假村，另外一间浴室则用灯光增加空间深度，并采用高级的旋涡式按摩浴缸。

在亚科斯森展区旁边是伊奈展区。其中展品包括：一体坐浴盆、高级马桶及盥洗盆。科学比例制作的热轧钢抵消掉了展厅密闭的感觉，而且金属与石头相结合的产品展示装置为产品增添了奢华感。

位置相对靠后且面积较大的美标展区分为浴室展示区及样品间。水盆及瓷器根据不同质地在不同的区域展示。浴室展示区在墙上展示了5种设计理念设计的浴缸，并用薄金属框架分隔开来。

■ 美标展区

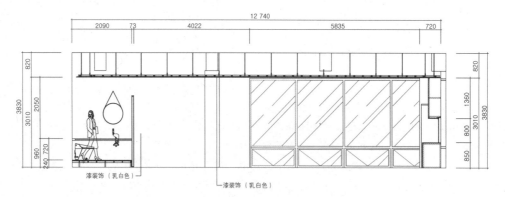

立面图R

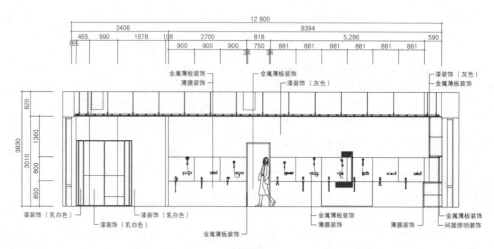

立面图S

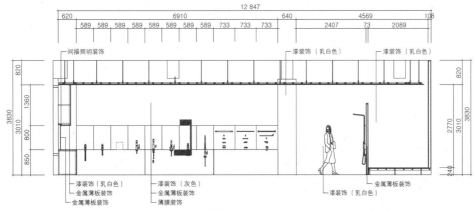

立面图T

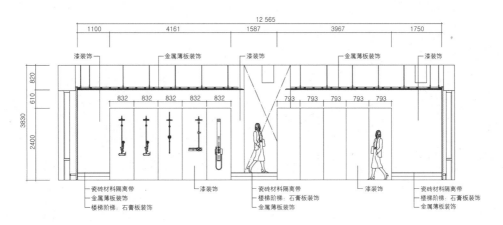

立面图U

"试衣间"

建筑面积：180平方米

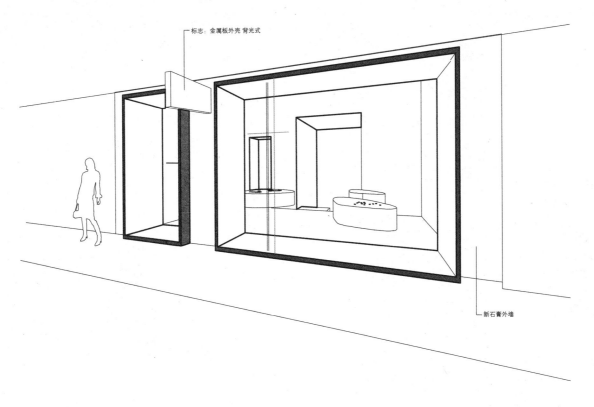

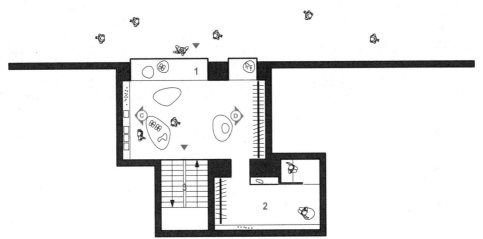

一层平面图

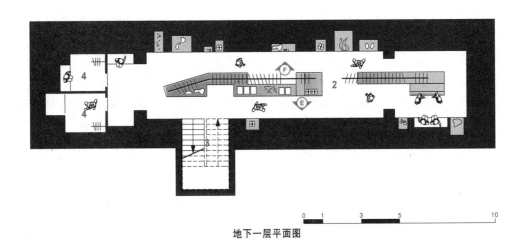

1. 入口
2. 展示区
3. 楼梯间
4. 试衣间

地下一层平面图

"试衣间"

An old, venerable seed store in the heart of Vienna's center closes for retirement reasons. The central location in the boutique district is desirable, although it has an unattractive configuration with a small ground floor area and a large seed storage space in the basement. The highly traversed Graben and Stephansplatz are in the direct vicinity.

Flooded with light, the pure white ground floor opens towards the street. This space is a stage and calling card at once. The discreet boudoir on the ground floor is connected to the show stage on the courtyard side. The central location of the stairs and their function as a connection between the two levels have a great significance. The stairs, visible from the Petersplatz, invite an exploration and lead to the basement calmly, naturally. With its resemblance to a tunnel, the underground sales area is conceived as a dark hull. Display cases that permit the light to enter the dark space are embedded in the walls and ceiling. A large room sculpture is in the middle, on which the current collection is presented.

位于维也纳市中心的一家古老而优美的种子商店因店主退休原因而休业。精品区的中心位置是极其抢手的，虽然它的面积较小，让人不甚满意，但是在地下室却拥有一个极大的种子储藏空间。它横穿格拉本大街，与斯蒂芬广场比邻。

洒满灯光，全白色的地板装饰一直通向大街。一楼的会客室通向院子边的展示台。位于店铺中心位置的楼梯担负着连接两层楼的重要任务。从斯蒂芬广场可以直接看到店中的楼梯。顾客可以以一种平静自然的方式顺着楼梯探索到地下室。地下室的设计类似于通道，这样的设计使得地下室的打折区给人一种暗色船体效果。展示箱嵌入天花板及墙面，使得灯光映入这个黑暗的空间中。房间中央是一个巨大的雕塑展台，上面陈列着最新的产品。

入口

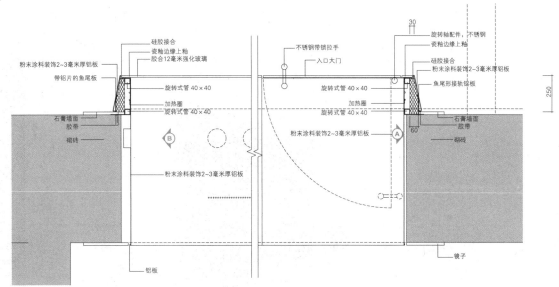

局部平面图

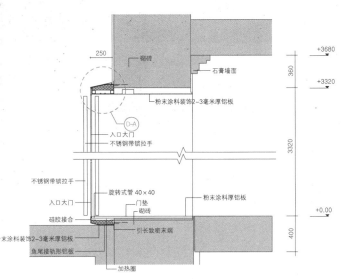

立面图A 　　　　　　　　细节图A

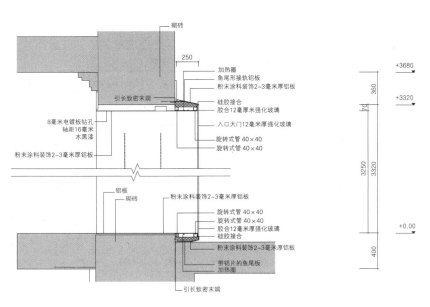

立面图B

■ 一楼展示区

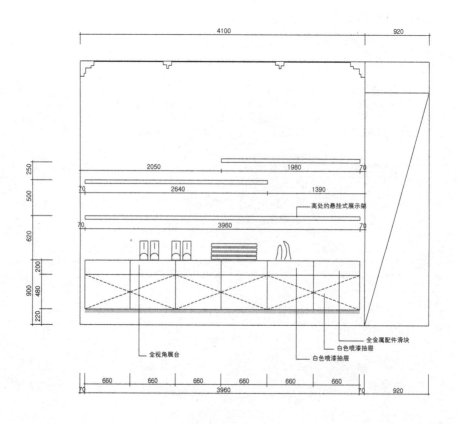

立面图C

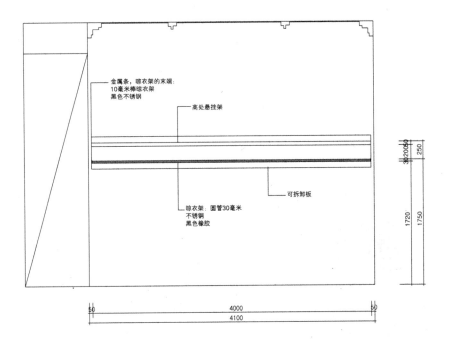

立面图D

■ 一楼展示柜

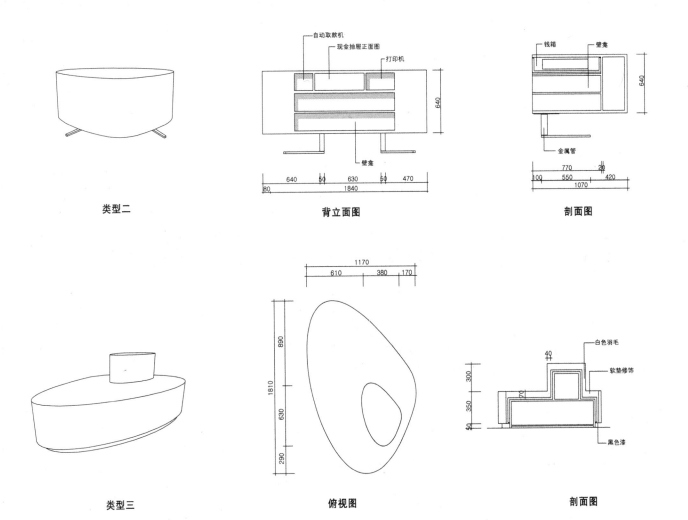

类型二　　　　　背立面图　　　　　剖面图

类型三　　　　　俯视图　　　　　剖面图

■ 楼梯

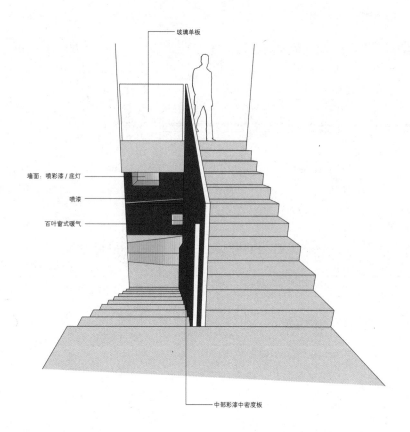

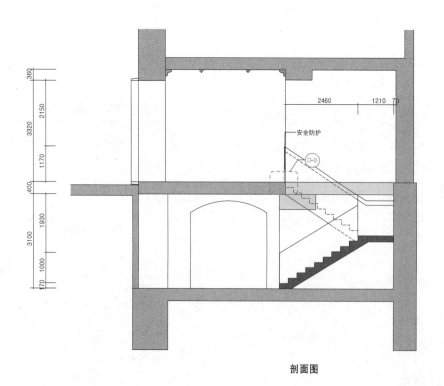

剖面图

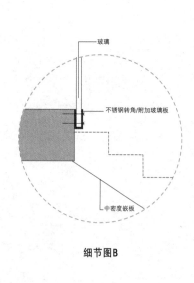

细节图B

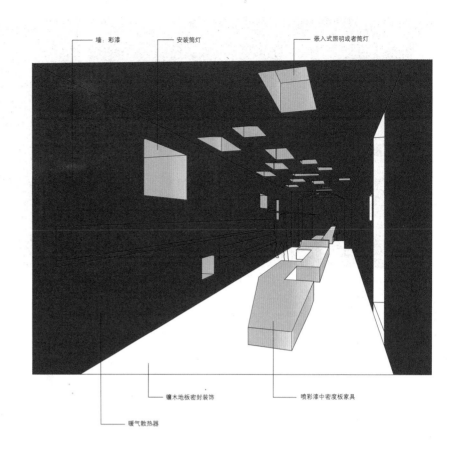

地下一层展示区

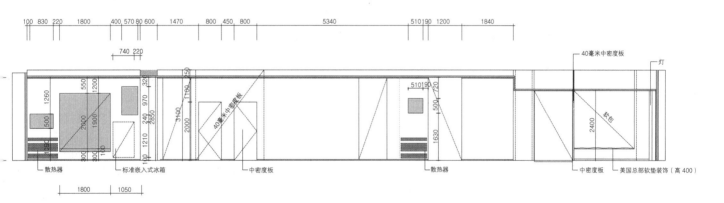

立面图 E

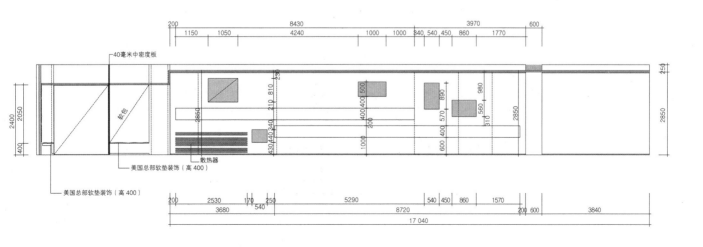

立面图 F

■ 地下一层展示区

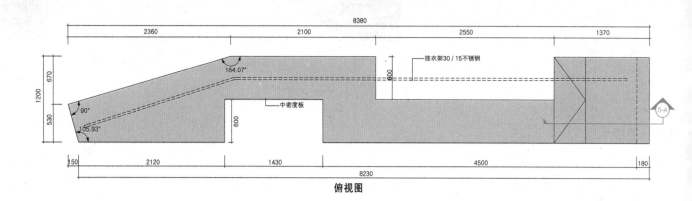

俯视图

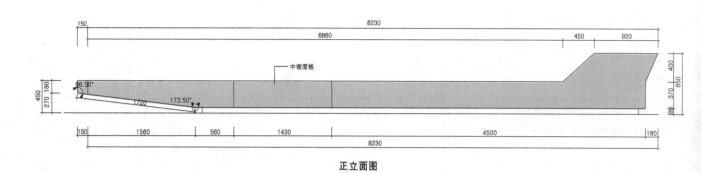

正立面图

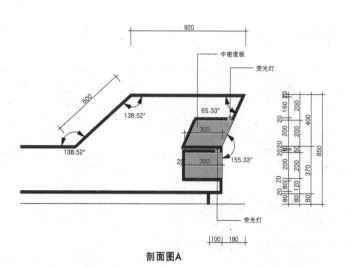

剖面图A

地下一层收银台

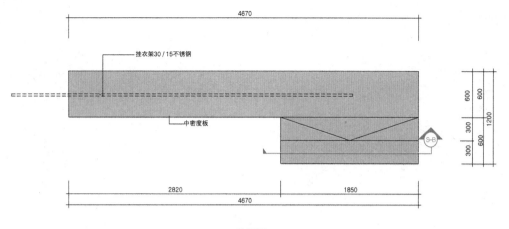

俯视图

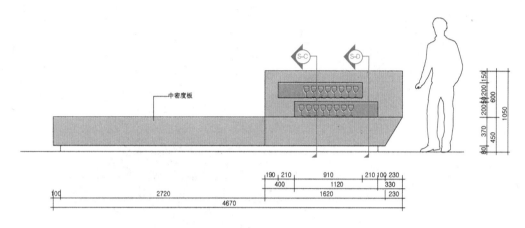

正立面图

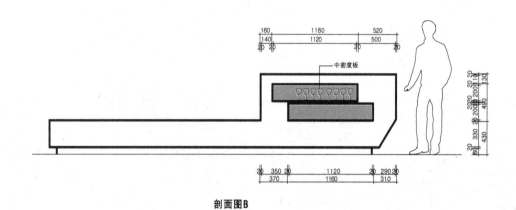

剖面图 B

剖面图 C   剖面图 D

迪普美发沙龙

建筑面积：120平方米

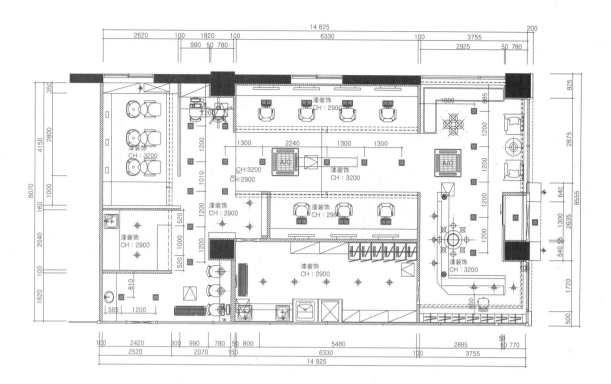

天花图

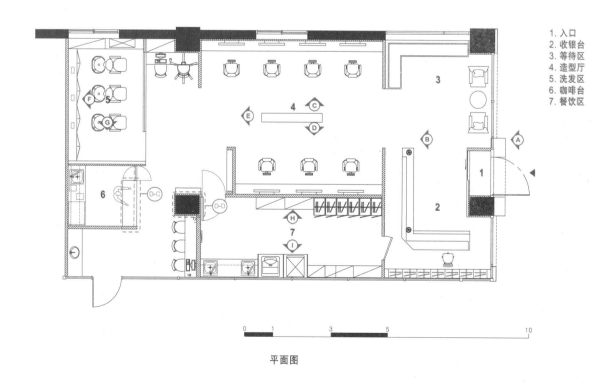

1. 入口
2. 收银台
3. 等待区
4. 造型厅
5. 洗发区
6. 咖啡台
7. 餐饮区

平面图

Being located in a not-yet-ordered new town, HAIR THE DEEP called for a noticeable element: the white door frame stands out between gray buildings. The pure white see-through curtains were drawn inside the windows on either side of the door to fulfill the aspect of decoration and spatial division. The contemporary and elegant interior themed with 'modern classic' is defined by four keywords: classic, service, luxury, and circulation. At the entrance, the white marble counter, black steel partition, and leather sofa present a luxurious ambience, while the marble-like concrete flooring brings a sense of unity and liberty. The classical molding on the wall was kept simple, and the mirrors framed in gold color accentuate and invigorate the space. The small cafe of a shop-in-shop style would satisfy the customers that spend quiet a long time waiting.

The efficient use of simple finish materials will offer the visitors a brilliant and sensuous spatial experience.

该项目使用了一些特别引人注意的元素，比如：灰色建筑物之间白色大门。每扇大门的窗户内部都有纯白色可透视的窗帘垂下，既满足了装饰的效果，又更好地划分了区域。极具当代艺术且优的室内设计通过经典、服务、奢华及循环四个重要元素，将店铺装修风格定位为"现代经典"。在入口处，白色大理石柜台、黑色属家具和皮质沙发显现出一种奢华的氛围，同时，大理石般的混土地面为店面添加了些许统一及自由的感觉。墙面上的经典嵌线饰设计将整体保持了一种简单的风格，而金色的镜框则让其更为机勃勃。小咖啡馆，这种店中店的设计可以更好地满足那些需要时间等待的顾客的需求。

这种极简且有效利用装饰材料的方式为顾客更好地提供了绝妙的间感。

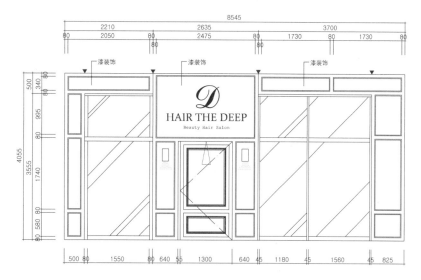

立面图A

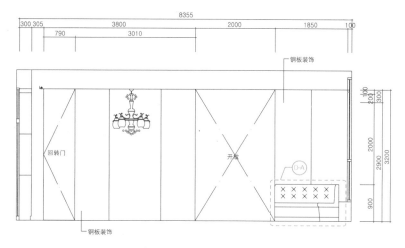

立面图B

■ 收银台

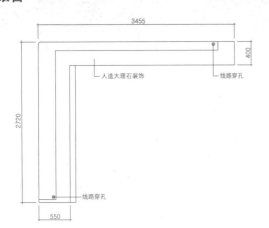

俯视图

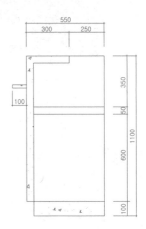

剖立面细节图

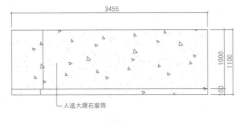

正立面图

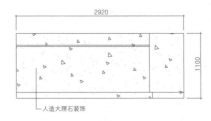

侧立面图

■ 细节图A——沙发等待区

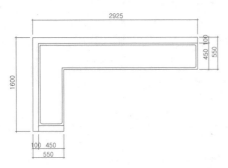

俯视图

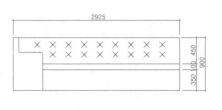

正立面图

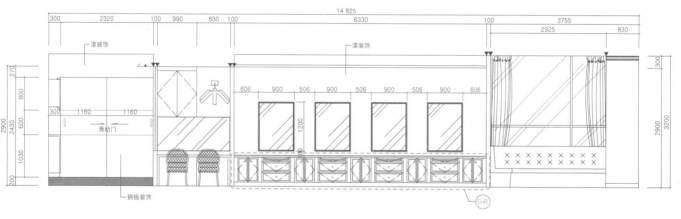

立面图C

细节图B——抽屉

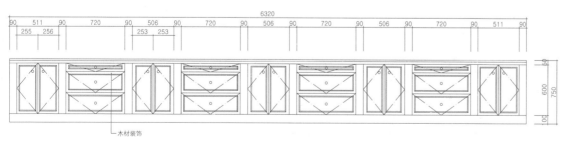 

正立面图　　　　剖面图

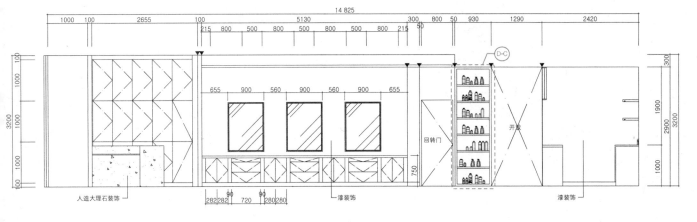

立面图D

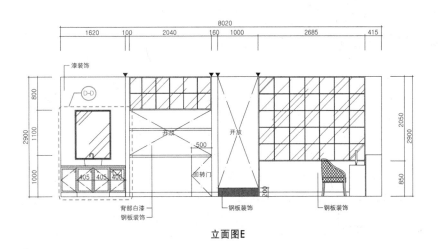

立面图E

细节图C　　　　　　细节图D

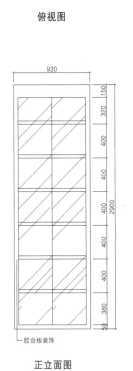

正立面图

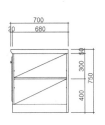

正立面图　　　　　侧立面图

■ 洗发区

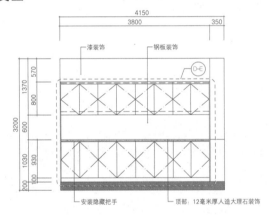

立面图F

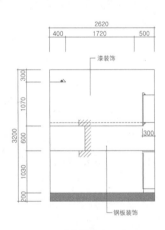

立面图G

■ 细节图E——架柜

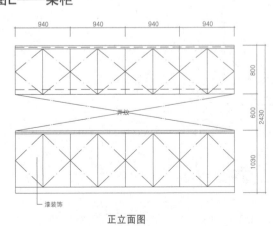

正立面图

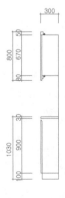

剖面图

餐具室

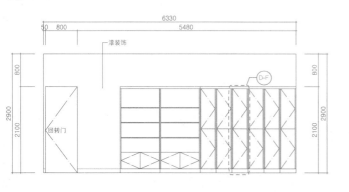

立面图H

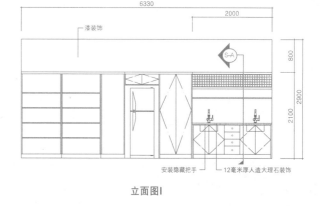

立面图I

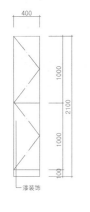

细节图F

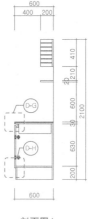

剖面图A

节点图G

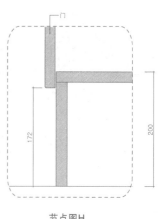

节点图H

| 金宝热轨书屋 | 建筑面积：927平方米

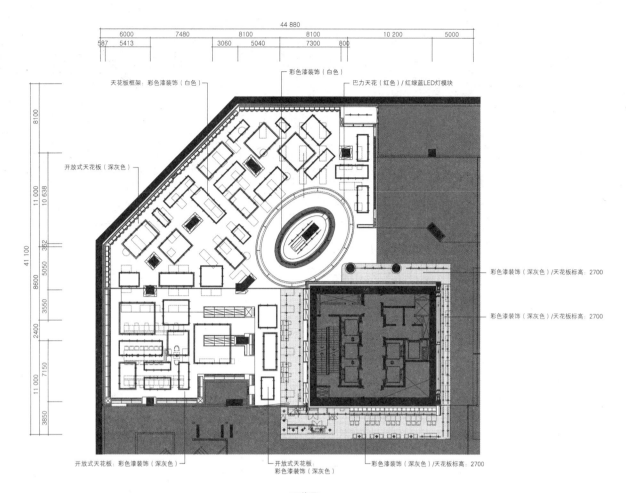

天花图

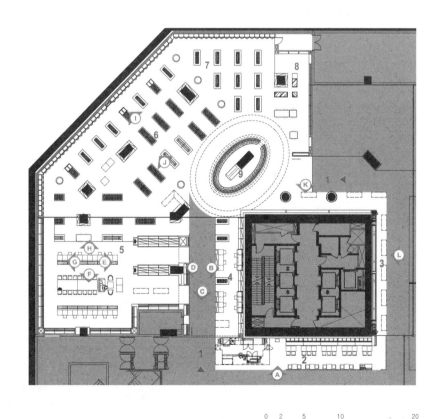

1. 入口
2. 咖啡厅
3. 促销区
4. 礼品图书区
5. 图书区
6. 文具区
7. 电子区
8. 高档钢笔区
9. 收银台及信息台

平面图

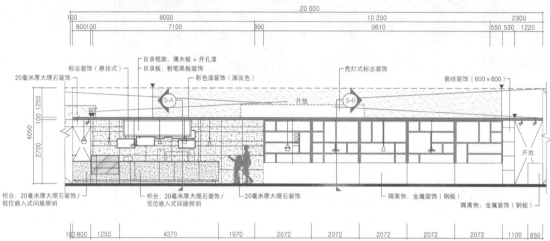

立面图A

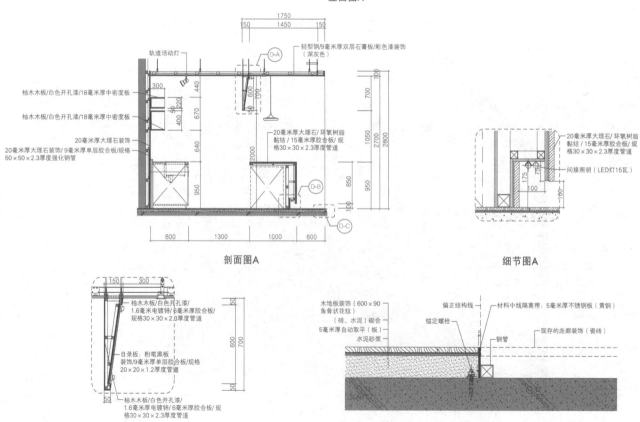

剖面图A

细节图A

细节图B

细节图C

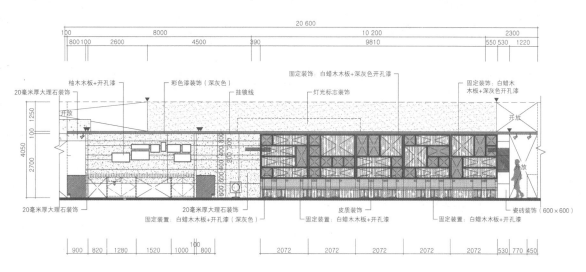

立面图A'

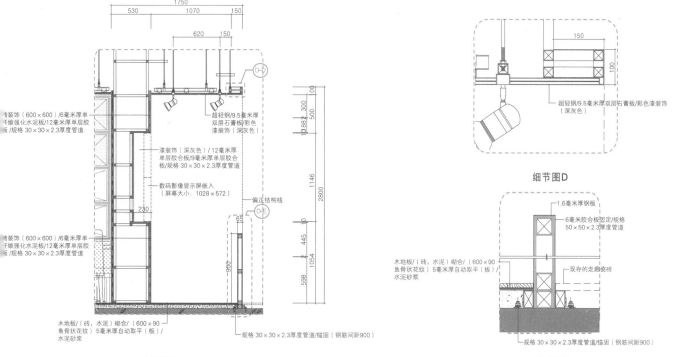

剖面图B

细节图D

细节图E

Located in D-Cube City, Kyobo Book Center has a complex cultural space with Dream Zone and Hot Tracks. Unlike other ones that focus on a bookstore, This branch is a functional space classified by theme under the concept of 'Designing for human use'; the designer analyzed human behavior as well as psychological and functional aspects based on human scale, range of motion, and character of behavior to apply them on the details of space, furniture and sign, creating an aesthetic and at the same time practical space.

Various products and relaxing areas provide a convenient shopping space for all ages and not just for 20s and 30s who used to be main customers. Considering the increase in sales of design products, Kyobo HotTracks offers an enjoyable cultural space themed with 'Environment providers who can tell and appreciate the excellent', 'The progressive in pop culture which is a measure of material development', and 'Provider of literature, arts, religion, scholarship, education, fashion, broadcasting, performance, and cinema'. Each section has a space for reading and listening to music, while the café at the entrance draws the visitors by offering a place for meeting and lounging.

位于 D-Cube City 的金宝图书中心，与 Dream Zone 及 Hot Tracks 共享着杂的文化空间。不同于其他只关注书本的书店，这间分店的设计题是"为人类使用而设计"。设计师在分析了人类行为模式、人类理因素及功能因素、活动范围及活动特点后，将这些应用到商店细节、家具和设计中，创造出一种极具美感，同时又充满实用性空间。

各种各样的产品和休息的空间为各个年龄段的人提供了方便的购空间，而不仅仅局限于那些过去常常被认为是主要顾客群的 20 岁30 岁的人群。考虑到设计品销量的增加，金宝热轨提供令人享受文化空间，主题分别是："有品位的环境提供者"、"物质发展标尺衡量" 和 "文学、艺术、宗教、学识、教育、时尚、广播、表演电影制作的提供者"。每个部分都有阅读室和听音乐的地方，同日在入口处也为顾客们提供了可以会面及休息的咖啡厅。

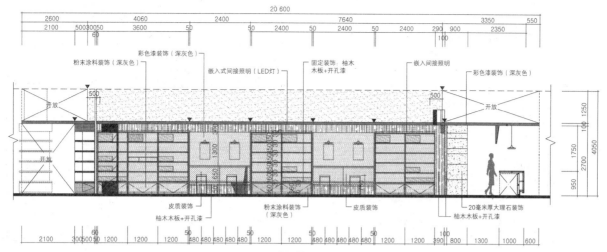

立面图B

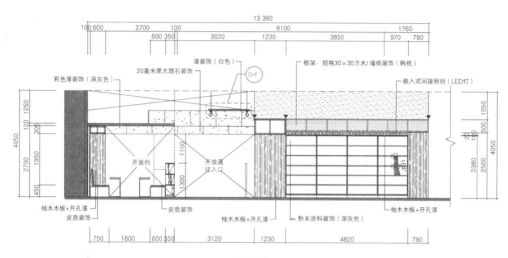

立面图C

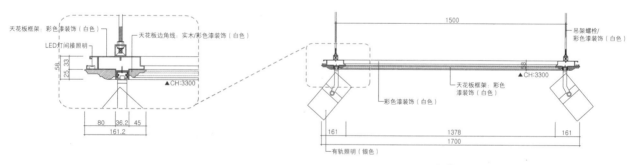

细节图F

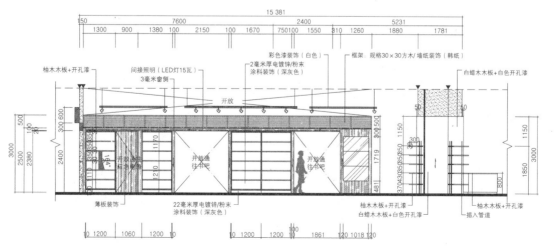

立面图D

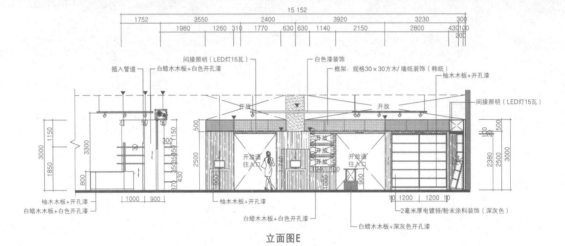

立面图E

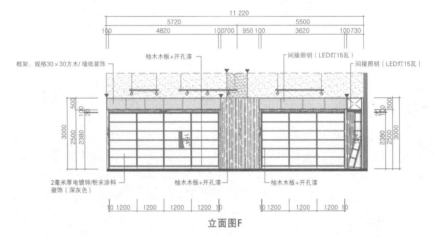

立面图F

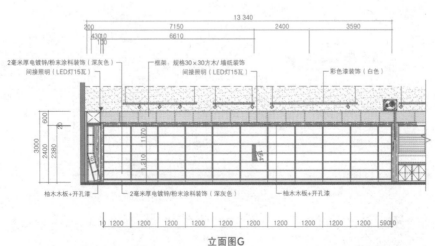

立面图G

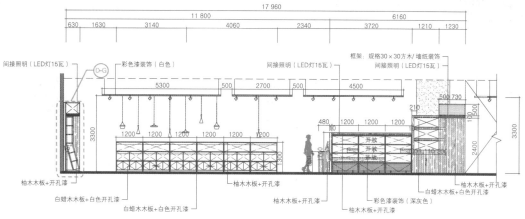

立面图 H

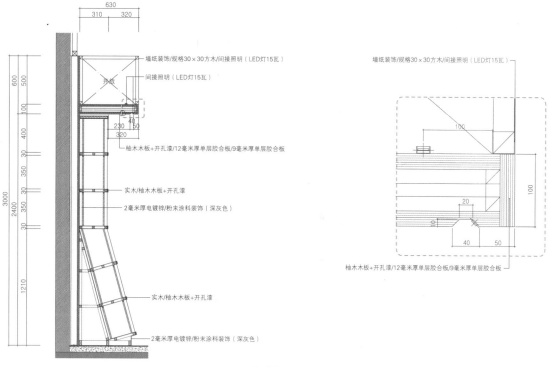

细节图 G

金宝热轨书屋 97

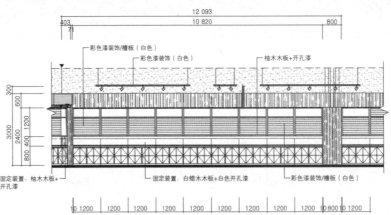

立面图I

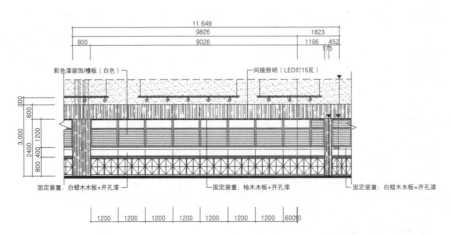

立面图I'

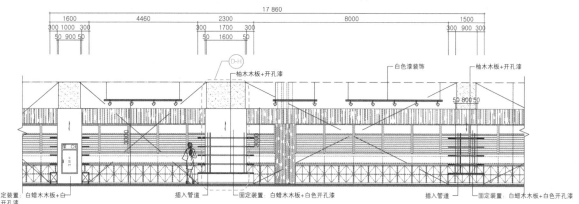

立面图J

细节图H——圆柱形展示柜

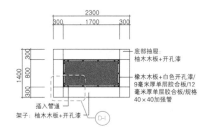

俯视图

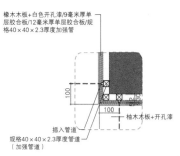

细节图I

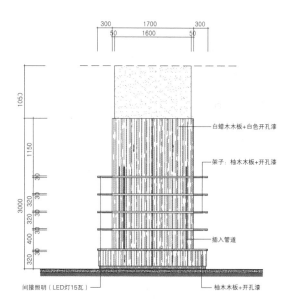

正立面图

金宝热轨书屋

■ 信息中心

顶视图

俯视图（剖面图）

剖面图C

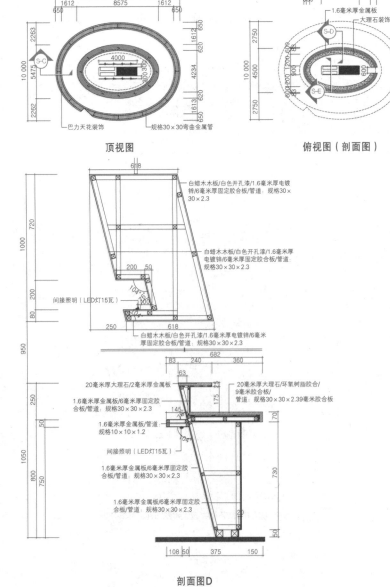

剖面图D

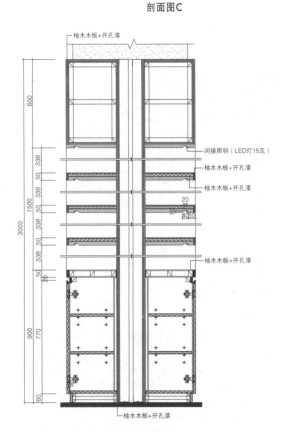

剖面图E

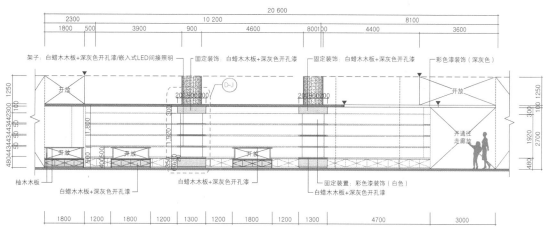

立面图K

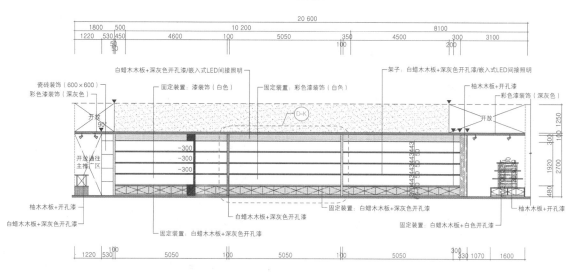

立面图K'

金宝热轨书屋

细节图J——圆柱形展柜

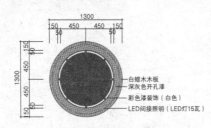

俯视图

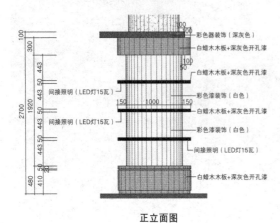

正立面图

剖面图

细节图J

细节图K——墙架

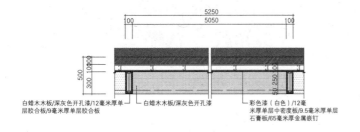

俯视图

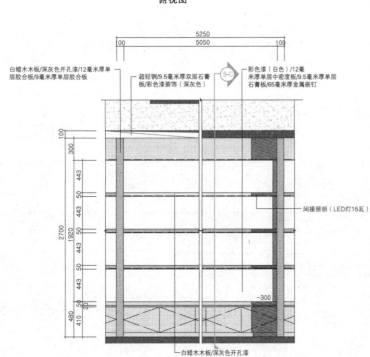

正立面图

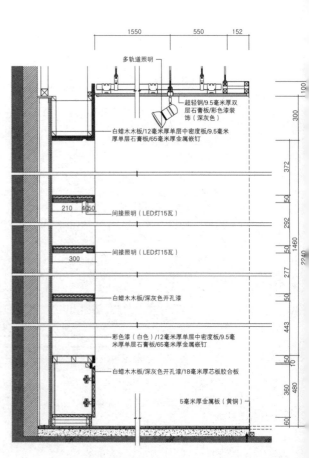

剖面图C

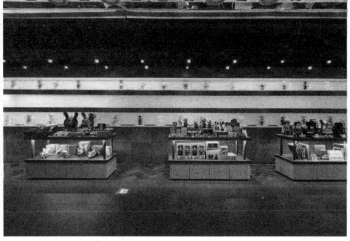
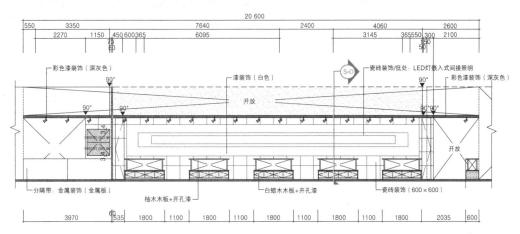

立面图L

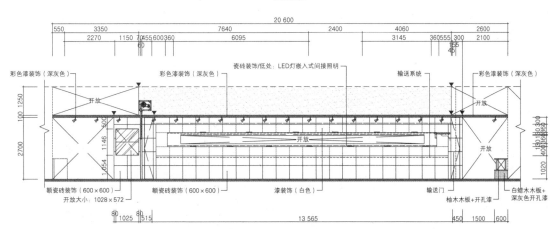

立面图L'

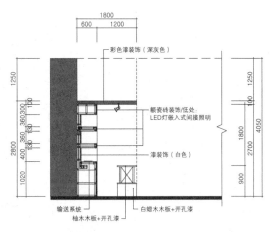

剖面图D

里科塔奶酪沙拉

建筑面积：35平方米

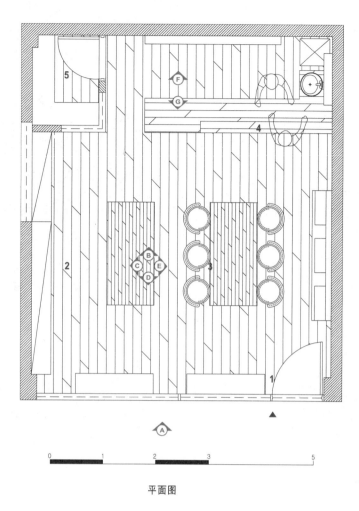

1. 入口
2. 展示区
3. 咖啡区
4. 收银台及厨房
5. 储藏室

平面图

外观

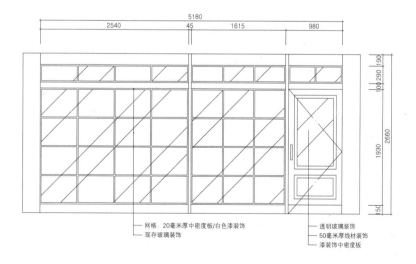

立面图A

里科塔奶酪沙拉

The store displays various styles of tableware like art works so as to differentiate itself from its competitors, which are classified by size and collection. The spacious interior is dotted by products; the central table can be a display area, tea table, and even as a workspace, thus adding a function of a cafe and an art studio.

The otherwise plain design was designed into a cafe-like store that offers a comfortable space for all ages.

该商店所展示的不同风格的餐具都宛如艺术品,而从众多竞争者中脱颖而出的原因是其所有展品均按照大小和收藏品的种类进行分类。商店使用不同的产品点缀宽敞的内部空间。中间餐桌可以用来当展示台、茶桌,或者是工作区,从而增添了咖啡吧和艺术室的功能性。

这个咖啡厅式的商店通过简单平实的设计为不同年龄段的人提供了舒适的环境。

收银台

立面图B

展示区

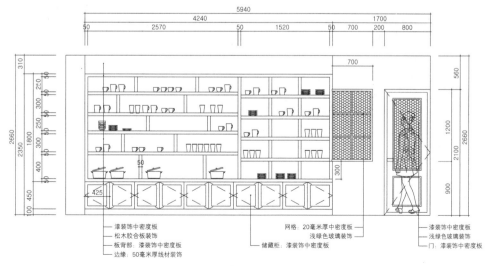

立面图C

里科塔奶酪沙拉

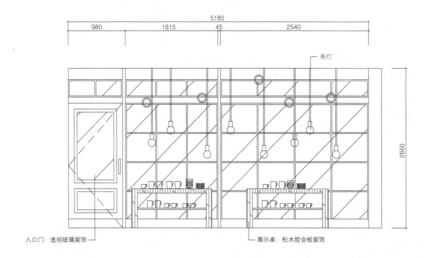

立面图D

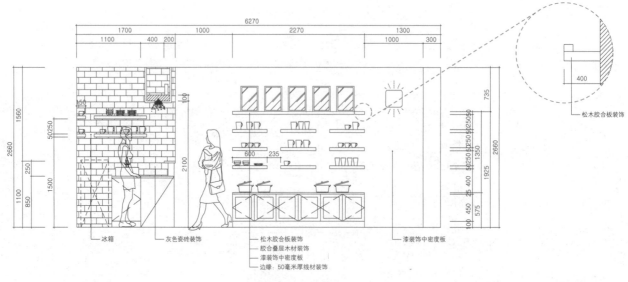

立面图E

厨房

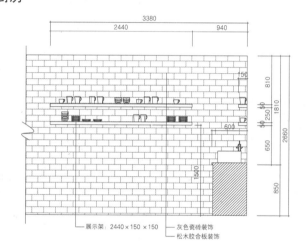

立面图F

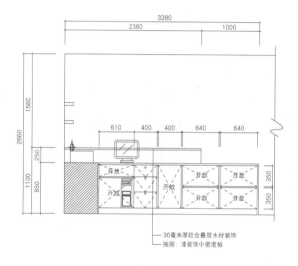

立面图G

展示桌

俯视图

正立面图　　侧立面图

■ 咖啡桌

俯视图

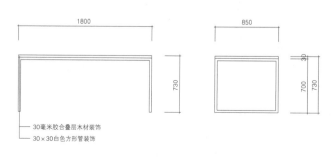

正立面图　　侧立面图

安利广场

建筑面积：2100平方米

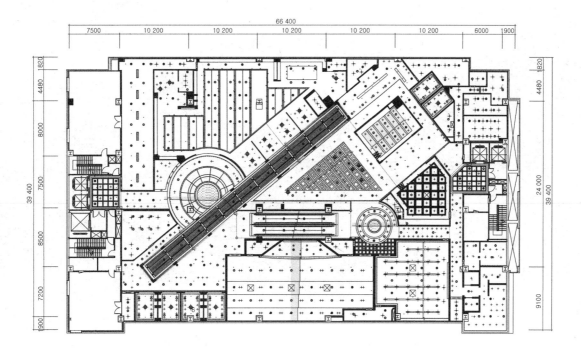

天花图

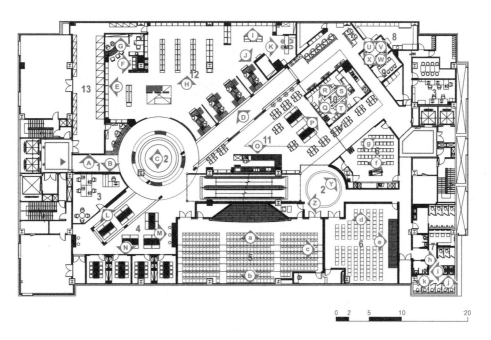

平面图

1. 入口
2. 广场
3. 办公室
4. 图书馆
5. 礼堂
6. 视觉中心
7. 厕所
8. 休息间
9. 会议室
10. 安利女王
11. 咖啡厅和休息室
12. 店铺
13. 储藏室

■ 入口

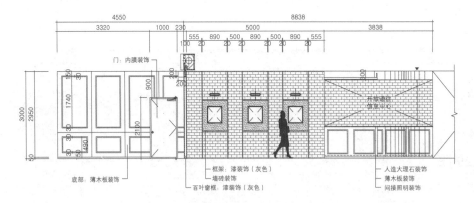

立面图A

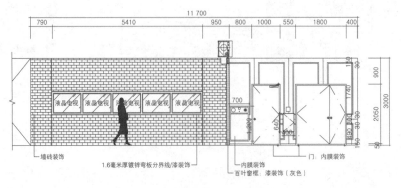

立面图B

店铺

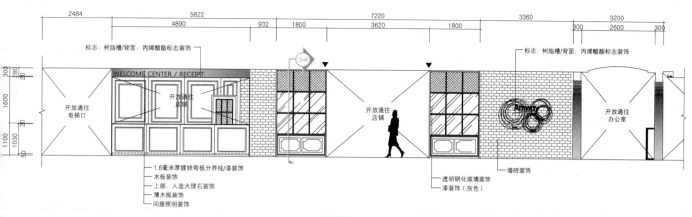

立面图C

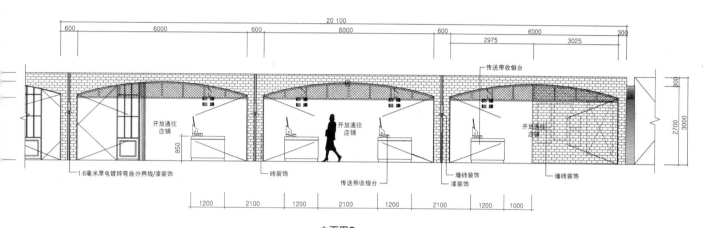

立面图D

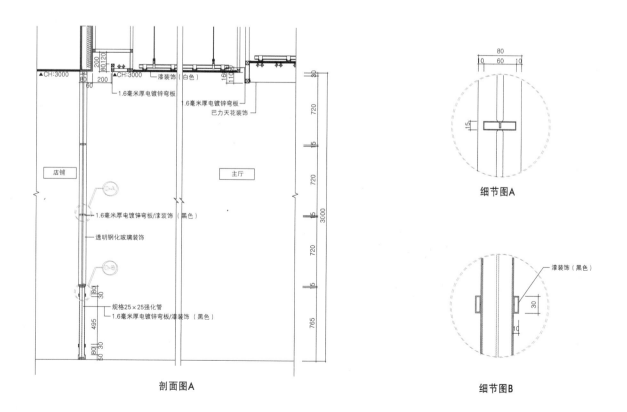

剖面图A

细节图A

细节图B

■ 店铺

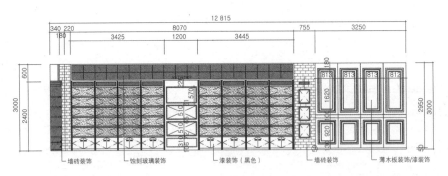

立面图E

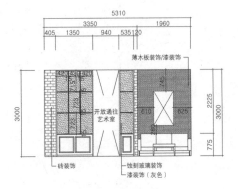

立面图F

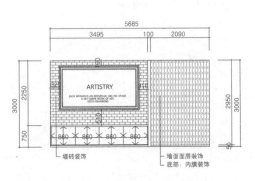

立面图G

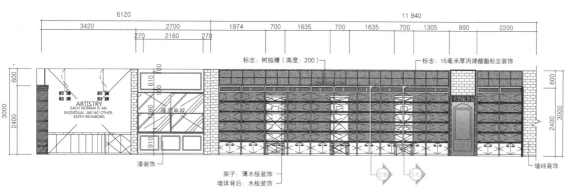

立面图H

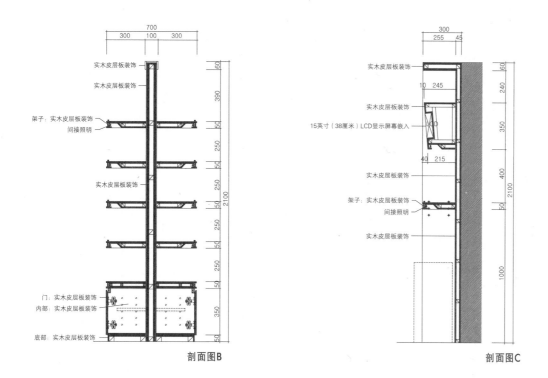

剖面图B   剖面图C

■ 纽崔莱

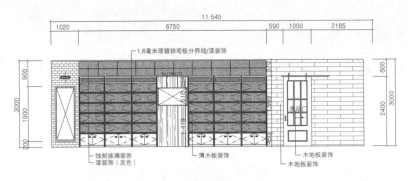

立面图 I

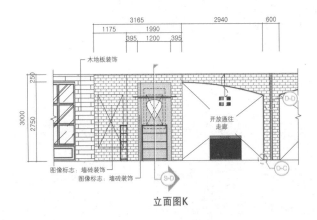

立面图 K

立面图 J

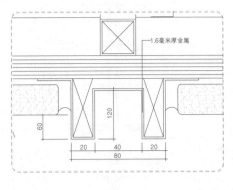

细节图 C　　　　　　　　　　细节图 D

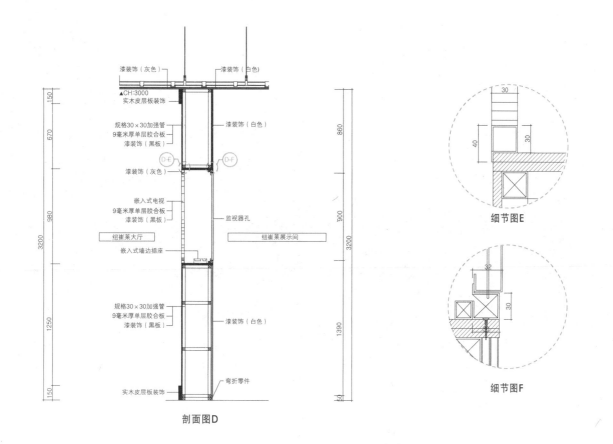

剖面图D

细节图E

细节图F

细节图G——办公桌

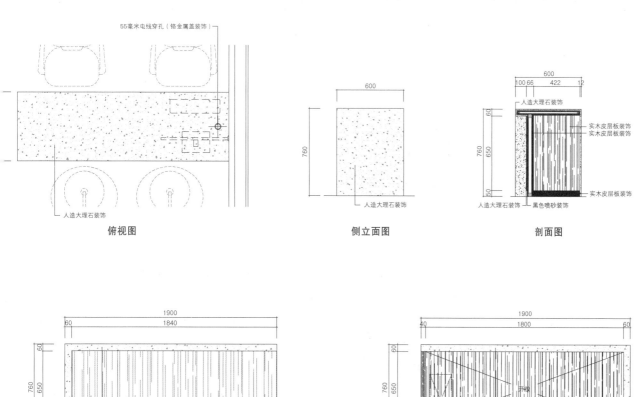

俯视图　　侧立面图　　剖面图

正立面图　　背立面图

■ 图书馆

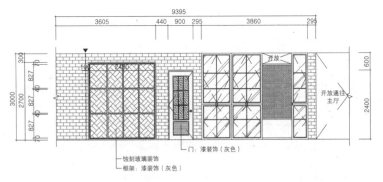

立面图L

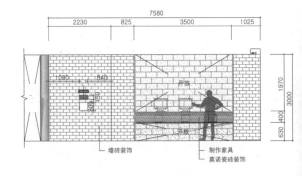

立面图M

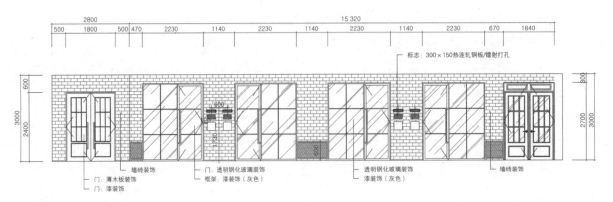

立面图N

This place is a 'total experience' space where sales of Amway products, health, beauty experience, lecture and seminar are available, which is designed for users to feel comfortable and relaxed.

The entire space is decorated with warm feeling using wood, brick and cozy lightings. At the center 50cm-long oblique arch passage is introduced to create the natural feeling like walking on an European street. The inside consists of square, market, lounge, auditorium, vision center, Amway queen, library, business room and conference room. Among them, the auditorium and 2 vision centers are special spaces which exist only in this place and do not exist in other branches. The auditorium with 300 seats is planned to hold a small scale performance and lecture by applying stair-like structure, theater-like chair and stage and dimming system. Vision Center 1 prepared for lecture emphasizes natural image with brick, sound-absorbing perforated panel and water board. Vision Center 2 is a multi-purpose space to be used as gallery, cuisine, lecture and party depending on whether or not the door is open by applying folding doors.

AMWAY PLAZA allows users to experience products more naturally in more comfortable and luxury environment and provides the best environment for shopping and business.

这是一个"全体验"的地方，出售安利产品，并提供健康和美容体验及讲座。这样的设计是为了让所有的客户感觉到舒适、放松。

整体空间通过使用木质、砖和暖光灯装饰出一种温暖的感觉。在中心位置建立了一个50厘米长的略倾斜的拱形走廊，创出一种随意行走于欧洲街头的感觉。内部设有广场、店铺、休闲区、礼堂、视觉中心、安利皇后产品区、图书馆、商务室及会议室。其中，礼堂和两座视觉中心是为该分店特别设置的，其他分店没有。礼堂以阶梯结构设置了300个座位，剧院样式的椅子、表演台和调光系统，主要用于开展小型表演及讲座。一号视觉中心主要用于授课，装饰手段更强调自然效果，大量使用环保砖、穿孔板吸声结构和防水板。二号视觉中心则为多功能厅，通过折叠门的开门实现用作展览馆、美食厅、演讲厅及聚会厅等功能。该中心的门可以通过使用折叠门达到开放及闭合的效果。

安利广场为顾客们提供了一种更自然、更舒适、更奢华的氛围，也为购物及商业洽谈提供了更好的环境。

休息室和咖啡厅

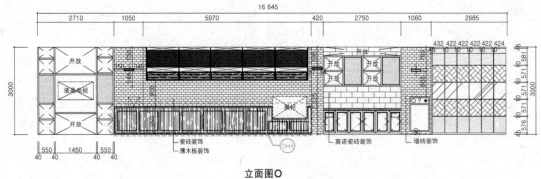

立面图O

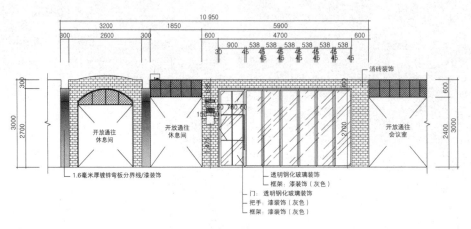

立面图P

细节图H——收银台

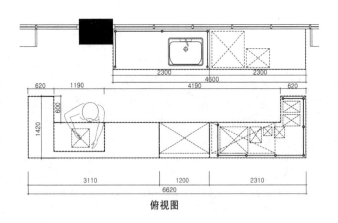

俯视图

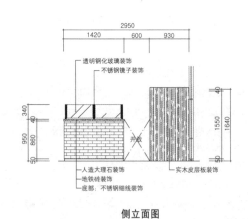

侧立面图

背立面图

内视图

安利广场

■ 安利皇后产品区

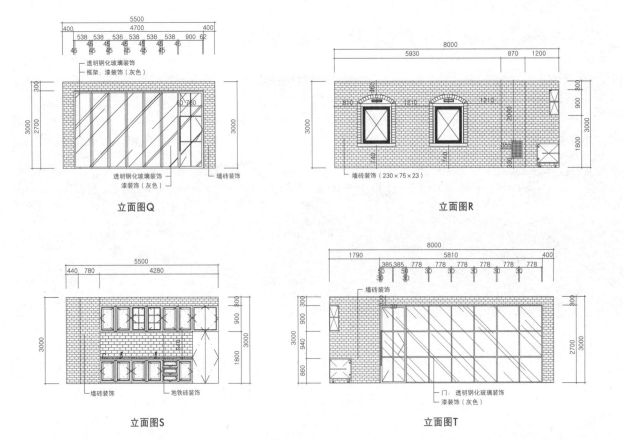

立面图Q

立面图R

立面图S

立面图T

会议室

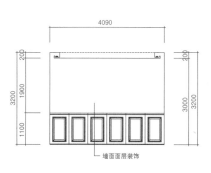

立面图U

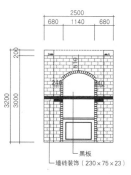

立面图V

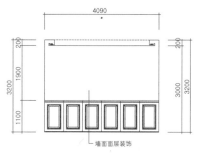

立面图W

立面图X

■ 礼堂大厅

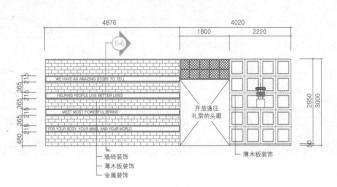

立面图Y

立面图Z

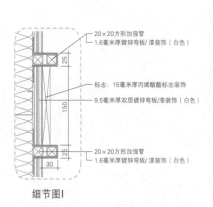

细节图I

剖面图E

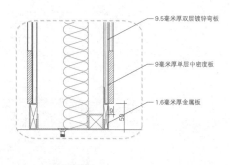

细节图J

礼堂

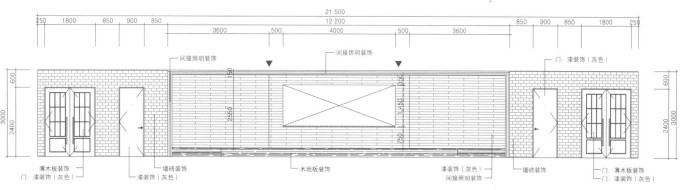

立面图a

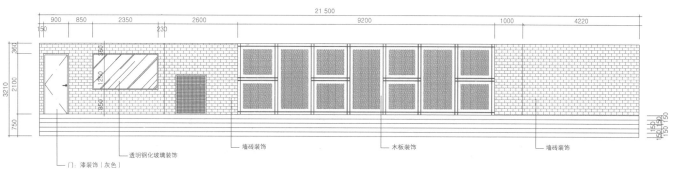

立面图b

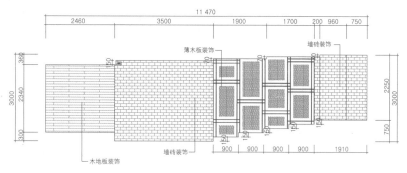

立面图c

■ 视觉中心

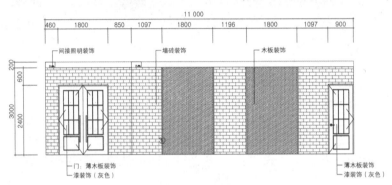

立面图d

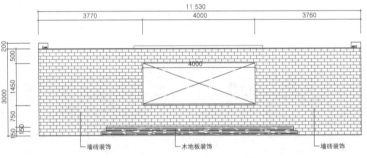

立面图e

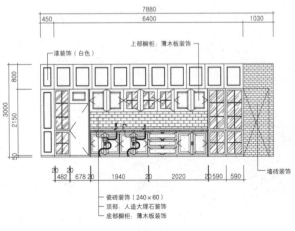

立面图f

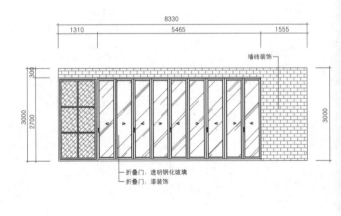

立面图g

厕所

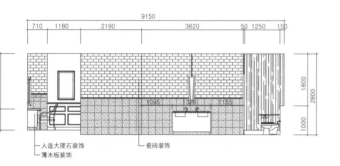

立面图h

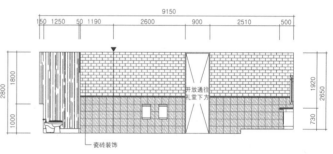

立面图i

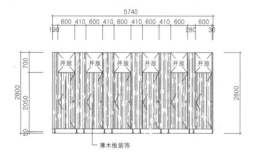

立面图j

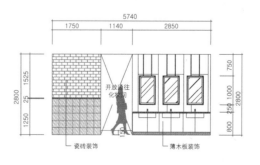

立面图k

住宅

永恒

帕克力拓住宅

H住宅

顶层公寓住宅

歌劳斯

马山住宅

| 永恒 | 建筑面积：398平方米

复式结构

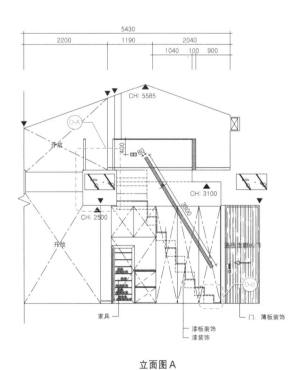

立面图A

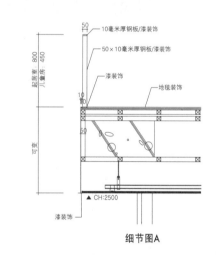

细节图A

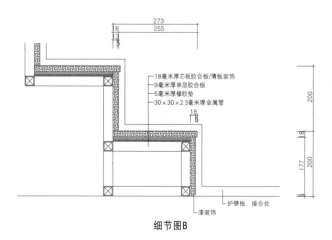

细节图B

Although living in a three-dimensional space, people are living their lives around flat and visual senses such as the color and the texture. Out of such two-dimensional design, this place displays a dynamic residential space which allows to experience a real sense of space.

By actively using the advantages that the household on the top floor has, the space is expanded to an optimum level within the scope that is allowed by the architecture framework, and the characteristics of the space are emphasized through various changes in the ceiling and the lighting. Also, it is neatly designed in a way that the furniture and the objets are emphasized over the finishing materials so that they can naturally harmonize with the space. Especially, for the lounge and the children's room, a multi-stage module is prepared using a high ceiling, which upgrades the function of the residential space to the next level through a harmony between public and private spaces while granting a fun element to a dull apartment structure. Also, various elements are added including the interesting shape of stairs and the red color finishing to complete a unique and comfortable residential space.

虽然人们生活在三维世界中，但生活中还是充满了平面的色彩和形的视觉效果。如此出挑的二维设计，使该生活空间充满动态感让人可以体验到一种空间的真实感。

通过有效利用顶层的优势，整个空间扩展到建筑框架可以达到的致水平。整体空间通过天花板及灯光效果的变化凸显了该设计的质。同时，相比于着重使用装饰材料，这种简洁的设计更强调家和装饰品的设计效果，从而达到整个空间的自然融合。特别的是休息室和儿童房采用了复式结构的设计和挑高天花板的方式，使住宅空间的功能性进一步提升。公共区域和私人空间的巧妙融合普通的公寓式结构增添了趣味性。另外，通过使用有趣形状的楼及红色装饰材料等多种元素，使得该独特且舒适的住宅空间更为整。

家庭房

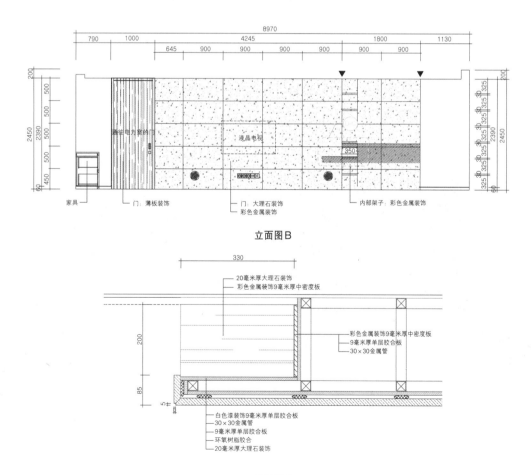

立面图B

细节图C

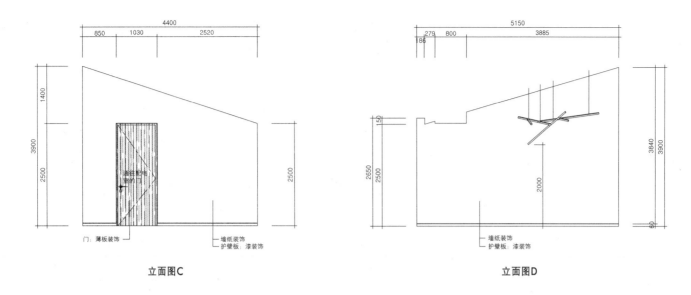

主卧——衣帽间镜子细节图

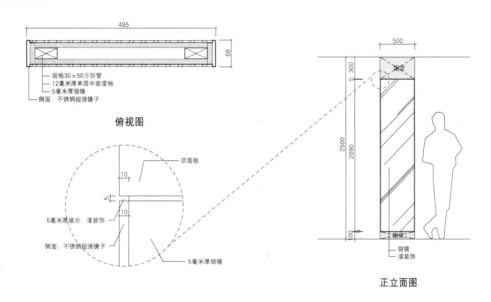

主卧——浴室

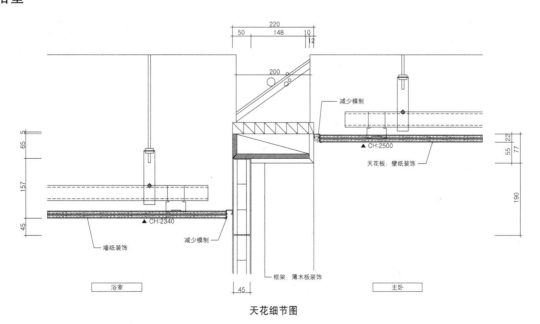

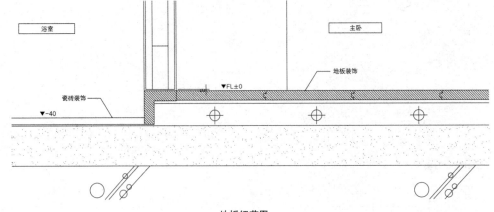

地板细节图

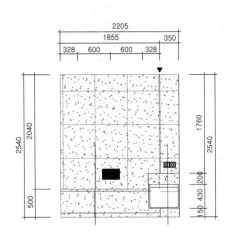

立面图E　　　　　　立面图F

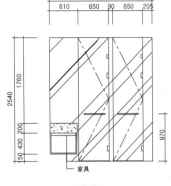 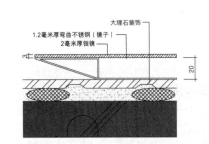

立面图G　　　　　　剖面图A

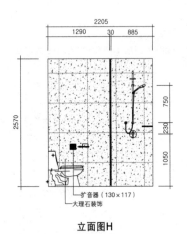 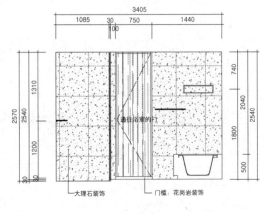

立面图H　　　　　　立面图I

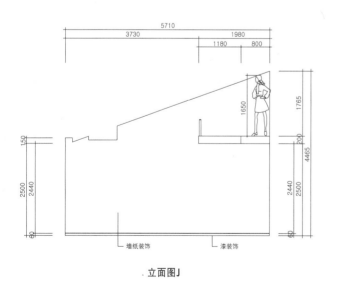

立面图J

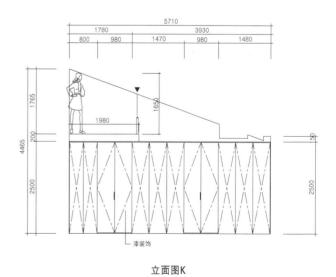

立面图K

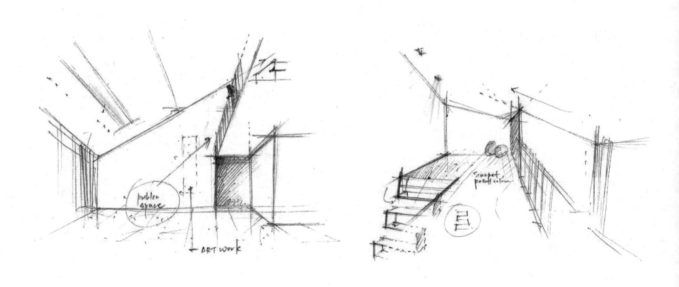

手绘图

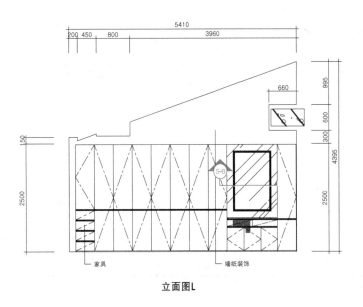

立面图L

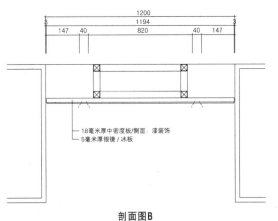

剖面图B

儿童房2——浴室

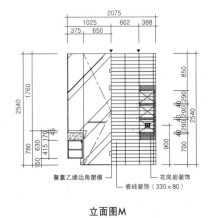

立面图M

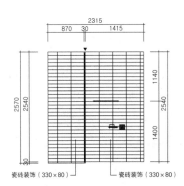

立面图N

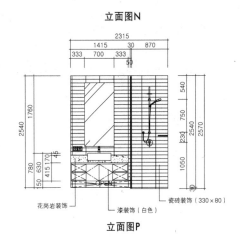

立面图O　　　　　　　立面图P

帕克力拓住宅

建筑面积：195平方米

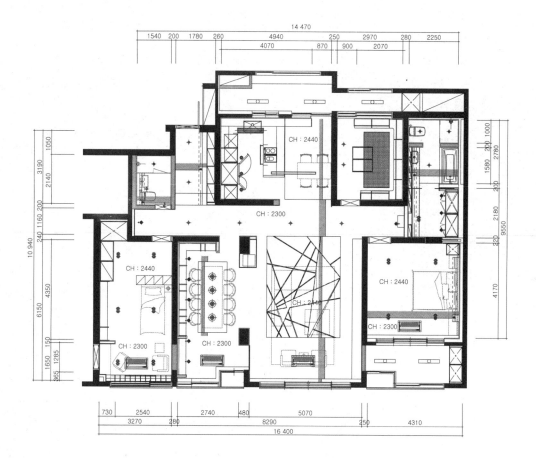

天花图

1. 入口
2. 房间
3. 艺术工作室
4. 起居室
5. 主卧
6. 阳台
7. 化妆间
8. 主卧浴室
9. 衣帽间
10. 厨房和餐厅
11. 杂物间

平面图

A residential space directly reflects the life of a family. Parkrio Residence suggests a customized residence for the client family consisting of a married couple and a 15 year old.

The existing four-bedroom space was renovated into two bedrooms, dressing room, and a second dining room, and the warm and calming finish materials bring a comfortable hotel-like atmosphere. The second dining room, which is rarely found in a housing unit, serves business gatherings that accompany family members, company dinner with the employees, and social meetings with neighbors or relatives. It is mainly used as a space of art activities such as displaying various works and objects, and listening to music. Two columns divide the living room and visually expand the space. Responding to the client's demand of a strong design element as seen in a commercial space, the designer introduced a divided pattern to the lighting box on the ceiling to spice up the living room.

Along with functionality, the customized design elements are added to create a sensuous residential environment.

一个住宅空间可以直接映射出一个家庭的生活质量。帕克力拓住为结婚15年以上的夫妻提供了一种定制的住宅空间。

之前四间卧室的空间被重新装潢成两个卧室、一个衣帽间和另一副餐厅。温暖且使人平静的建筑材料给整体空间增添了一种舒适酒店式的环境。第二个餐厅的设置，在单元住宅中很少出现。可作有家人陪伴的商业会面，与员工的公司聚餐及亲属邻居的社交面。更主要的是，这个餐厅还可以用来举办艺术活动，比如展示术品或者享受音乐。两列式结构将起居室分成两部分，并从视觉扩大了空间。应客户的要求，设计师引用一种经常在商业设计中到的强烈的设计元素，在天花板灯箱引入分块图案设计，这种设为起居室带来了活力。

在保持实用性的同时，这种定制的设计元素使整体住宅环境极具感。

入口

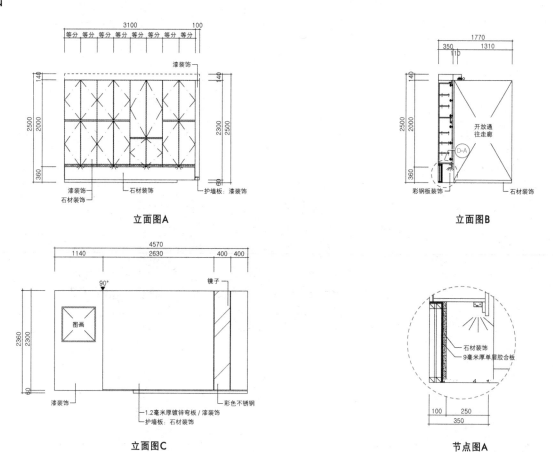

立面图A

立面图B

立面图C

节点图A

走廊

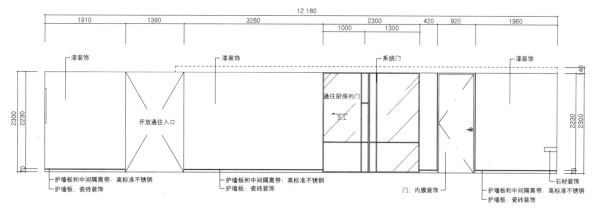

立面图D

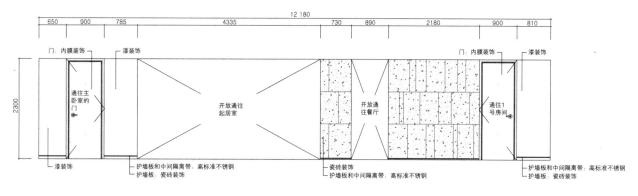

立面图E

■ 起居室

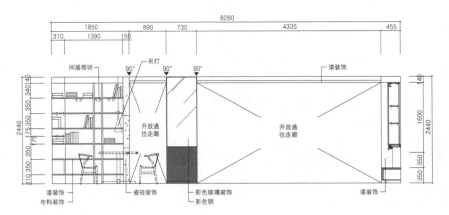

立面图F

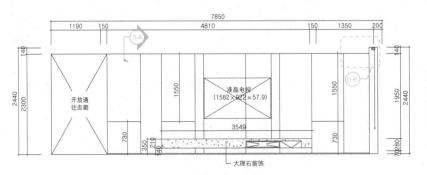

立面图G

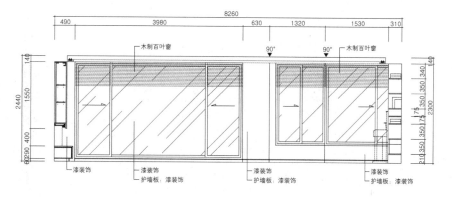

立面图H

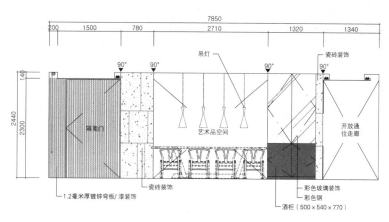

立面图I

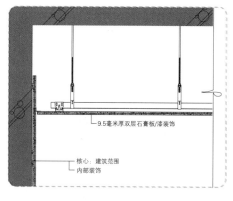

剖面图A

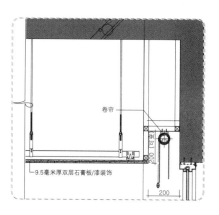

细节图B

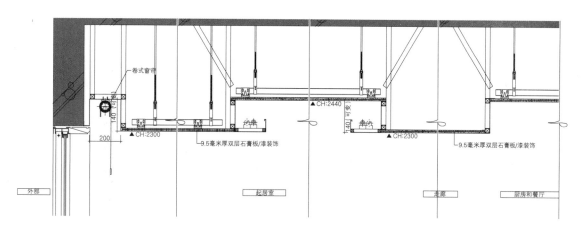

细节图C

■ 艺术室

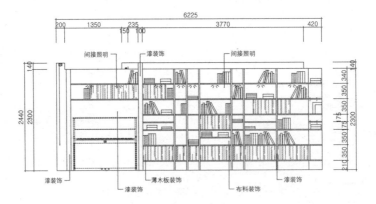

立面图J

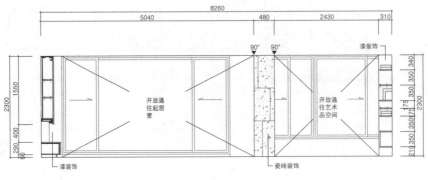

立面图K

厨房和餐厅

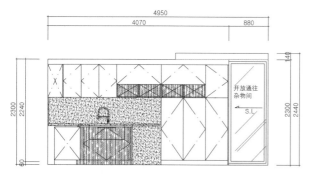

立面图L

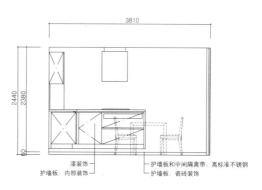

立面图M

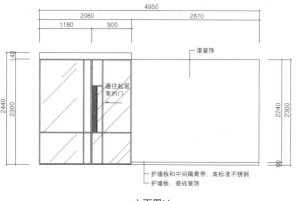

立面图N

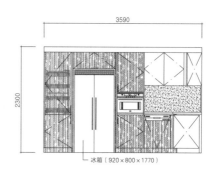

立面图O

■ 主卧

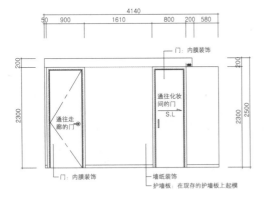

立面图P

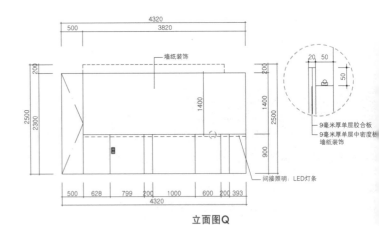

立面图Q

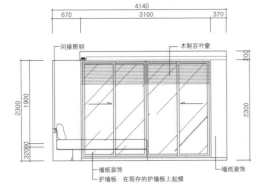

立面图R

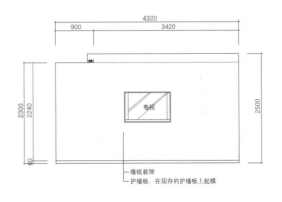

立面图S

化妆间和浴室

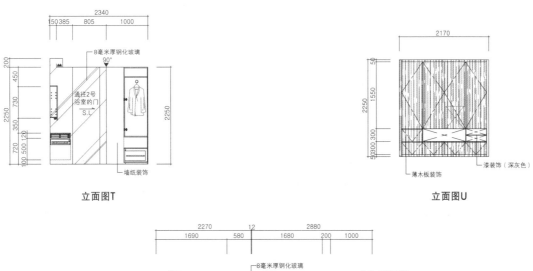

立面图T

立面图U

立面图V

|H住宅| 建筑面积：225平方米

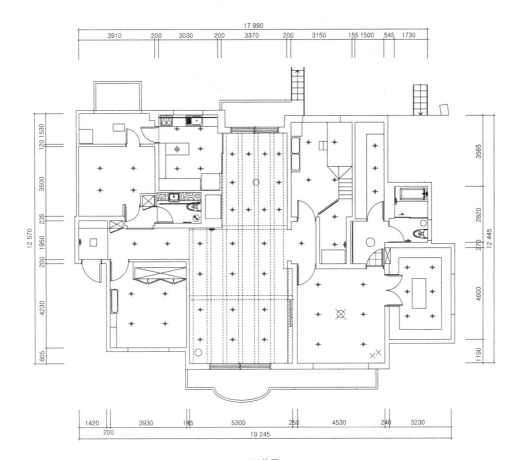

天花图

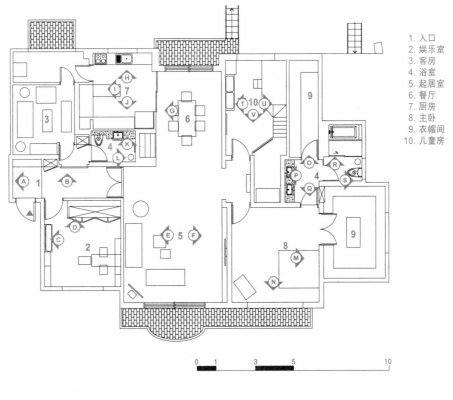

1. 入口
2. 娱乐室
3. 客房
4. 浴室
5. 起居室
6. 餐厅
7. 厨房
8. 主卧
9. 衣帽间
10. 儿童房

平面图

■ 入口

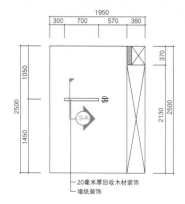

立面图A

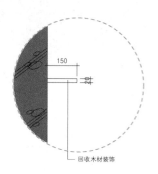

剖面图A

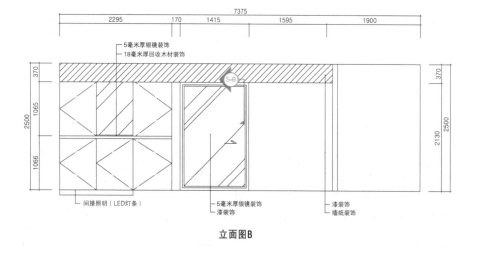

立面图B

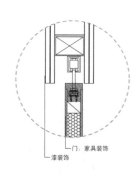

剖面图B

■ 娱乐室

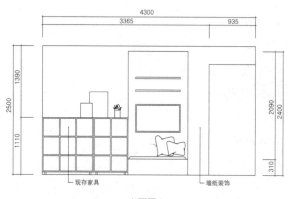

立面图C

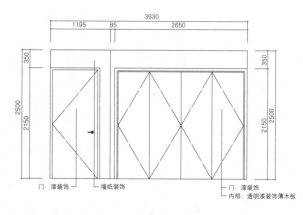

立面图D

起居室

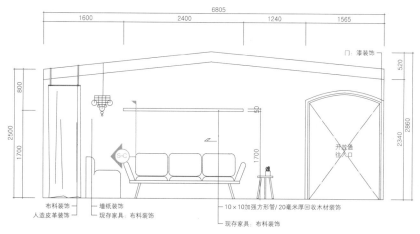

立面图E

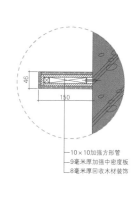

剖面图C

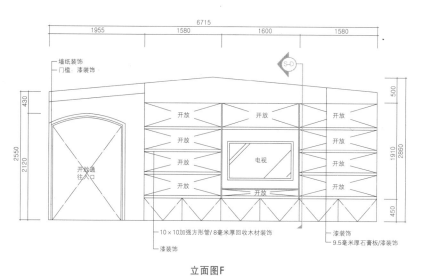

立面图F

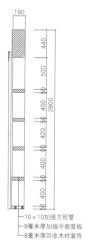

剖面图D

■ 餐厅

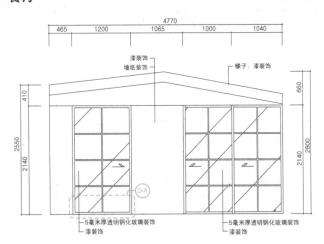

立面图G

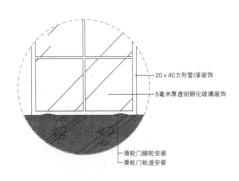

细节图A

■ 厨房

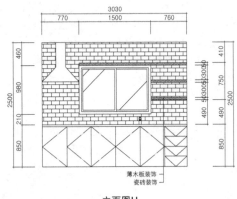

立面图H

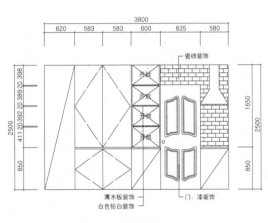

立面图I

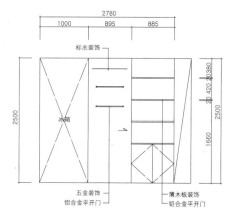

立面图 J

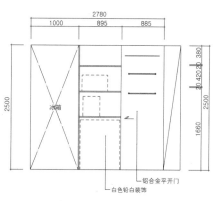

立面图 J'

浴室

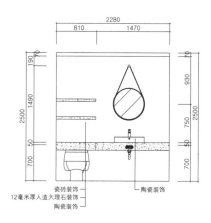

立面图 K

立面图 L

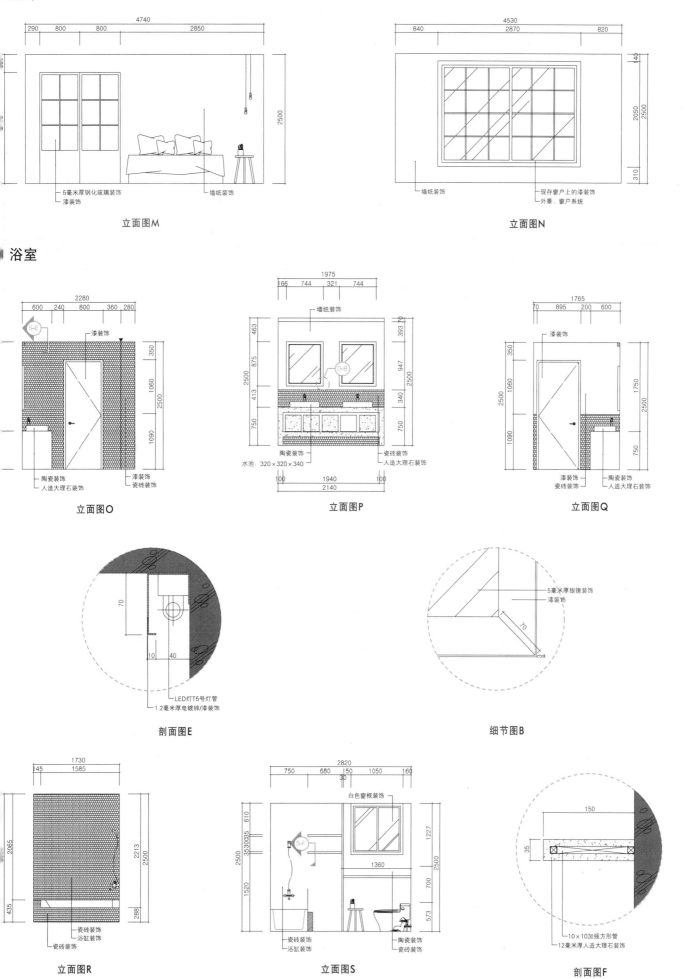

■ 儿童房

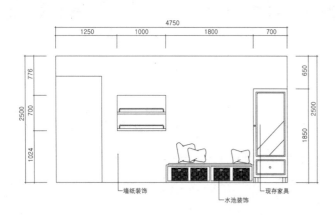

立面图T

立面图U

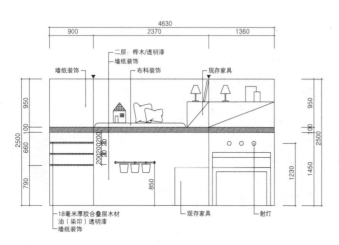

立面图V

立面图V'

e old villa was renovated into a new house with a warm and bright ambience. spite of its spacious 231-square-meter area, the interior emit a stifled atmo- ere. Thus the designer reconstructed the overall layout and added white finish to existing classical design to present a sense of brightness.

e small entrance has a built-in boot cupboard and a huge mirror. The long corri- separates the living room from the dining room. In the living room, the sliding ors conceal the TV set and small equipments. The wall design was kept to a min- um, while the bookshelves made of old wood deliver a vintage feel. Opposite the ng room is a dining room, in which vintage wood furniture exude a sense of rmth. The tall cabinet, wood shelves and sliding doors give the kitchen a neat d practical look. Lying beyond the arch at the end of the corridor are the master droom and the kid's room. Against the gray background, the furniture, fabric, and iced windows bring coziness into the master bedroom. The dressing room in the k of the bedroom has an island type cabinet, hangers, and shelves to ensure a ficient storage space for the owner who works as a fashion designer. The adjoin- bathroom is enlivened by the mosaic tiles in cheerful colors. The kid's room ing the bedroom has a duplex structure for sleeping and playing. The play room family and a guestroom meet various demands.

sic design and a sense of vintage are combined to create a sensuous residential ce.

这幢老房子被重新装修成了一个新的温暖而明亮的别墅。尽管房子面积有231平方米，可是内部氛围仍是让人感到窒息。因此，设计师将多余的分层全部拆除，并且重新为这个经典的设计增加白色装饰，使得整个空间充满了明亮感。

在相对狭小的入口处有一个嵌入式鞋柜和一面大镜子。长廊将起居室与餐厅分隔开来。在起居室，滑动门将电视和一些小型家具隐藏起来。墙面设计保持极简风格，同时，老木书架的使用为房子增添了一种复古的感觉。餐厅位于起居室的对面，大量仿旧家具的使用使餐厅散发着温暖的气氛。高橱柜、木质架子和滑动门的设计使厨房看起来整洁且实用。在拱形走廊的最末端，是主卧与儿童房。在灰色背景的映衬下，家具、布料和花格窗的使用使主卧更显安逸。在主卧背面的衣帽间使用了中岛型设计的橱柜、各种衣架和衣柜，确保即使是时尚设计师户主也有足够的空间可以使用。各种活泼颜色的马赛克的使用，使得与卧室相邻的浴室充满活力。卧室对面的儿童房则使用了双层结构设计，将休息与玩乐结合起来。家庭娱乐室和客房满足了其他各种需求。

极简设计及复古效果融合使用创造极具美感的住宅空间。

顶层公寓住宅

建筑面积：250平方米

1. 入口
2. 主卧
3. 衣帽间
4. 浴室
5. 餐厅
6. 起居室
7. 露台
8. 厨房
9. 办公室和客房
10. 洗衣房

平面图

起居室

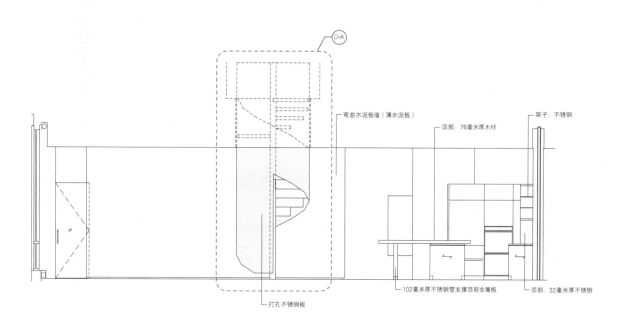

立面图A

顶层公寓住宅

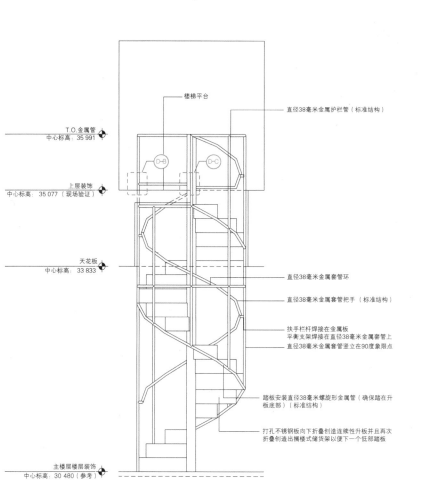

细节图A

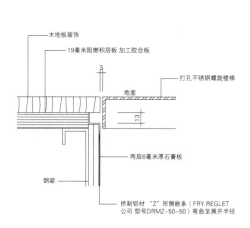

细节图B

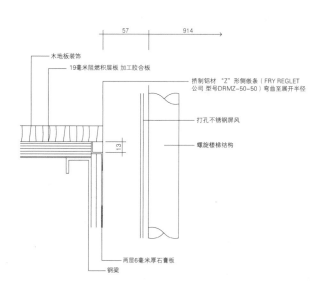

细节图C

顶层公寓住宅　163

楼梯

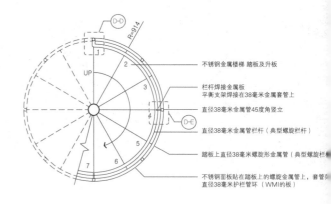

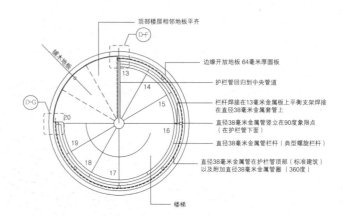

局部平面图

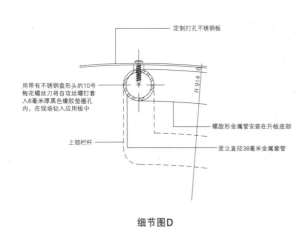

细节图D

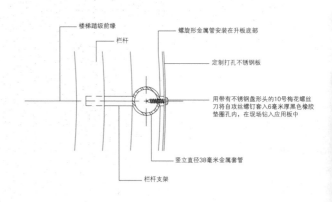

细节图E

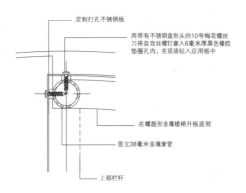

细节图F

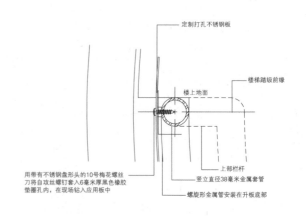

细节图G

  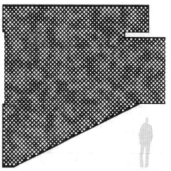

形象来源
由于工程临近明尼阿波里斯市五大湖流域,设计理念源自于水的隐喻(如图所示)。

操作过程
视觉效果设计是为了创造一种体现原来水形象的半色调纹理。这个马赛克图像是参考了三个直径不变的圆(89毫米、76毫米、63毫米)。

结果
该马赛克纹理在上述平坦的不锈钢板上层叠展现。该电子图像曾用于引导电脑激光切割不锈钢板。

楼梯板处理

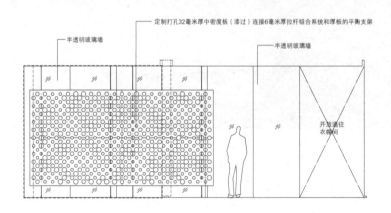

立面图B

his sixth floor penthouse overlooks to city lakes, the Uptown retail district and the city skyline beyond. Designed for a young professional, the space is shaped by distinguishing e private and public realms through sculptural spatial gestures. Upon entry, a curved wall white marble dust plaster pulls one into the space and delineates the boundary of the rivate master suite. The master bedroom space is screened from the entry by a translucent ass wall layered with a perforated veil creating optical dynamics and movement. This nctions to privatize the master suite, while still allowing light to filter through the space to e entry. Suspended cabinet elements of Australian Walnut float opposite the curved white all and Walnut floors lead one into the living room and kitchen spaces.

 custom perforated stainless steel shroud surrounds a spiral stair that leads to a roof eck and garden space above, creating a daylit lantern within the center of the space. The oncept for the stair began with the metaphor of water as a connection to the chain of city kes. An image of water was abstracted into a series of pixels that were translated into a eries of varying perforations, creating a dynamic pattern cut out of curved stainless steel anels. The result creates a sensory exciting path of movement and light, allowing the ser to move up and down through dramatic shadow patterns that change with the osition of the sun, transforming the light within the space.

e kitchen is composed of Cherry and translucent glass cabinets with stainless steel helves and countertops creating a progressive, modern backdrop to the interior edge of e living space. The powder room draws light through translucent glass, nestled behind e kitchen. Lines of light within, and suspended from the ceiling extend through the space ward the glass perimeter, defining a graphic counterpoint to the natural light from the erimeter full height glass. Within the master suite a freestanding Burlington stone athroom mass creates solidity and privacy while separating the bedroom area from the ath and dressing spa ces. The curved wall creates a walk-in dressing space as a fine outique within the suite. The suspended screen acts as art within the master bedroom hile filtering the light from the full height windows which open to the city beyond. The uest suite and office is located behind the pale blue wall of the kitchen through a sliding translucent glass panel. Natural light reaches the interior spaces of the dressing room and bath over partial height walls and clerestory glass.

这个位于第六层的顶层公寓可以鸟瞰城市湖景、上层商业区和城市地平线。为了迎合年轻专业人士的喜好，整体空间将私人与公共区域明显地分隔开来。在入口处，延伸出一面弯曲的白色大理石粉墙，勾勒出私密主卧套房的界限。从入口处竖起的透明玻璃墙完美地隐藏了主卧空间，而透孔的光影帷幕创造出一种光的动态美。这既起到了保护主卧套房隐私的功能，同时又使光洒满整个空间。由澳大利亚核木制造的悬浮橱柜悬挂在弯曲的白墙对面，而核木地面将人们带入起居室和厨房。

定制的带孔不锈钢将环形楼梯包裹起来，将人们带到屋顶和花园，并在整个空间的中央造找出一种灯笼的效果。楼梯设计的灵感来源于水，暗喻了公寓与城市湖景的关系。水的形象抽象分解成一系列像素点，从而转化为一系列不同的圆形图案，通过弯曲的不锈钢板创造出动态的花纹。这令人兴奋的灯光和移动路径的设计使得整体的动态光影效果随着太阳的位移和灯光的变化而产生新的图案。

厨房是由樱桃色和透明的玻璃橱柜构成，不锈钢支架及工作台为起居空间形成前卫的、现代的背景。光穿过透明的玻璃照射到位于厨房后方的衣帽间。光线从天花板上延伸下来，穿透整个空间到玻璃边缘，与透过全景窗而来的自然光形成图形般的交界点。主卧套房内，在卧室、浴室和衣帽间中间设立了一个由伯灵顿石墙构建的隐私空间。曲面墙创造出的可出入的换衣空间形成了该公寓内的精品换衣间。悬浮的屏幕在主卧像艺术品一样，同时，还可以过滤从面对城市地平线的全景窗透入的光。客房和书房位于厨房的浅蓝色墙体背后，可通过滑动式半透明的玻璃板进入。自然光透过玻璃窗照射进来，使整个衣帽间和浴室格外明亮。

| 歌劳斯 | 建筑面积：180平方米

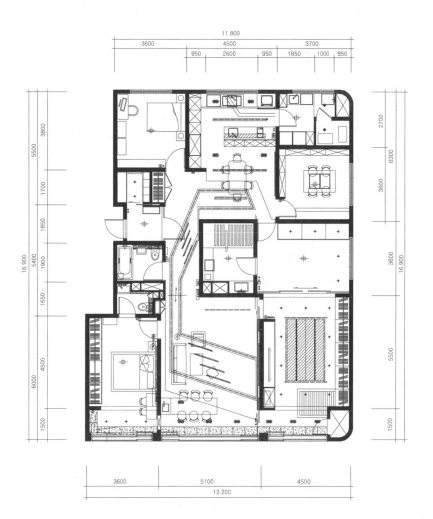

天花图

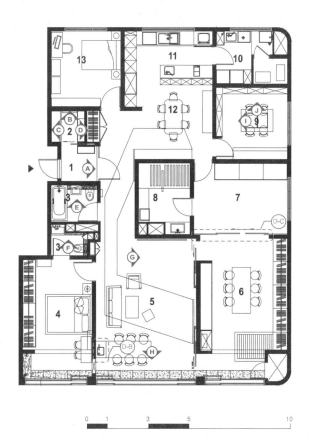

1. 入口
2. 鞋架
3. 浴室
4. 卧室
5. 起居室
6. 画室（公共区）
7. 画室（私人区）
8. 储藏室
9. 图书室
10. 杂物间
11. 厨房
12. 餐厅
13. 客房

平面图

歌劳斯 169

The client, who is both a painter and a collector, had sought out a space in which she could not only invite her guests to show her paintings, but also live comfortably in the environment that houses her tools and work. Hence the main goal of the project was to merge life with art, as well as the public with the private.

The new design scheme aimed to incorporate the functions of an art gallery/atelier and those of an apartment. The proportions of the existing apartment space had presented some practical difficulties: in contrast with the typical LDK-type apartments, which integrate the living room, the dining room, and the kitchen, the client's apartment had a C-shape wall in the center which separated the kitchen and the living room. The client complained of the resulting separation of her life patterns in the two areas of the apartment.

Instead of seeing this wall as an obstacle and destroying it altogether, the AnLstudio used it as a gallery wall while incorporating the use of geometrical lighting structures throughout the ceiling to suggest a linkage between the living room and kitchen. These fixtures were installed to provide the displayed elements with three different lighting sources-direct, indirect, and spot.

The gallery wall is visible from several key locations of the apartment: the entrance, the kitchen, the dining room, and the living room, while the lighting structures offer a 'flow' through the entire space and create a hybrid of the exhibition and living areas with a visual continuity.

该客户是画家，同时也是收藏家，她想寻求一个既可以邀请客人参观她的画作，又可以居住的舒适的工作生活空间。因此，该项程的主要目的就是要融合生活与艺术，达到公开与隐私的平衡。

新的设计理念要将艺术展览馆和公寓的功能性相结合。然而现存公寓空间已经出现了一些实用性的问题：与普遍存在的单卧室单厅单厨房的一体公寓不同的是，该客户的公寓中有一面C形墙将居室与厨房分隔开来。客户对于该设计在公寓内造成的分离式生模式非常不满。

AnL工作室将墙体当作画廊的展示墙，而不是当作一个障碍物将其部拆除掉。同时，与天花板上的几何形灯相结合，形成了起居室厨房之间的连接体。这些固定装置的安装与直接、间接照明还有光灯三种灯源的结合，为整体空间提供了更好的展示效果。

从公寓的几个重要位置都可以看到走廊的墙壁，比如：入口、厨房餐厅和起居室。同时，照明效果使得整体结构呈现一种"流动"效果，打造出展品展示与日常生活相融合的视觉连贯效果。

入口

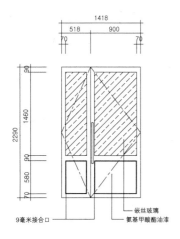

立面图A

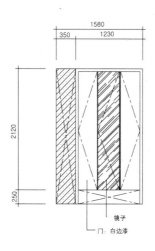

立面图B

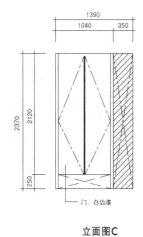

立面图C

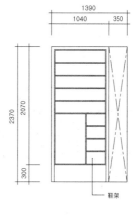

立面图D

浴室

立面图E

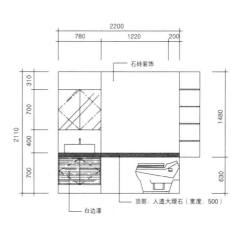

立面图F

■ 起居室

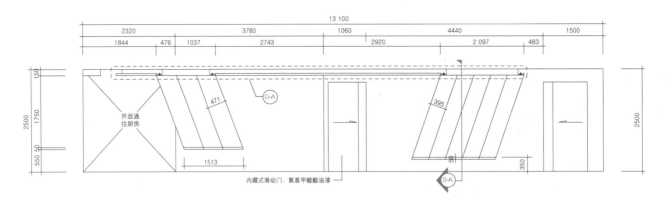

立面图G

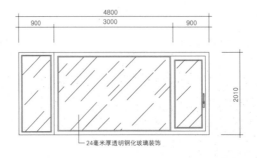

立面图H

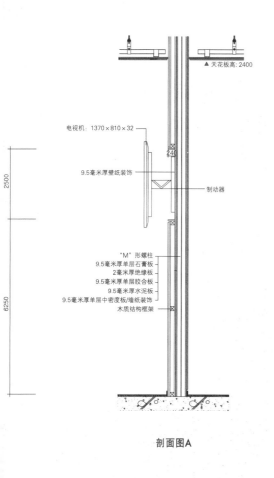

剖面图A

细节图A——天花板结构

■ 细节图B——家用吧台水槽

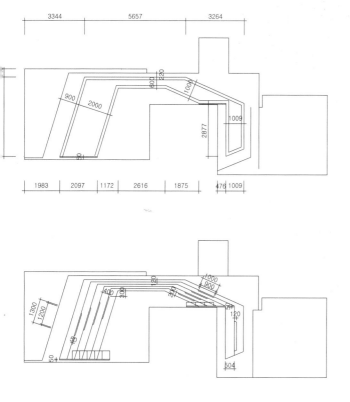

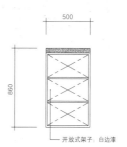

侧立面图

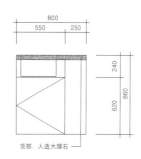

正立面图

■ 细节图C——滑动门

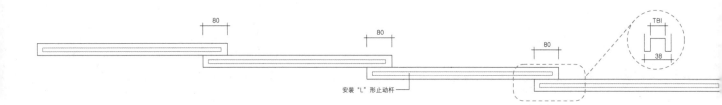

俯视图

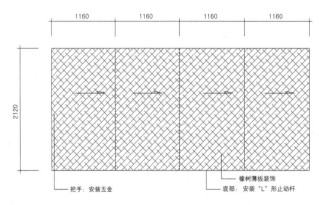

正立面图

图书馆

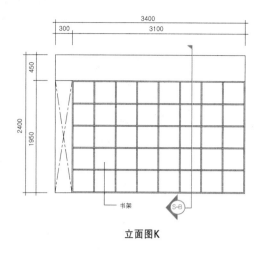

立面图K

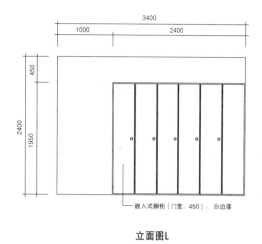

立面图L

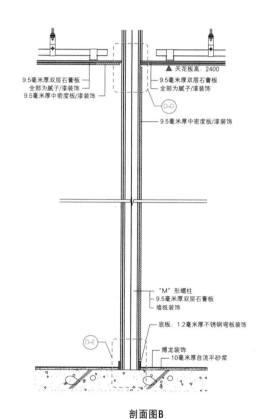

剖面图B

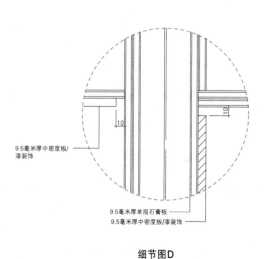

细节图D

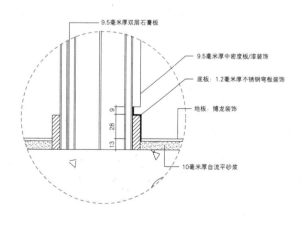

细节图E

马山住宅

建筑面积：609平方米

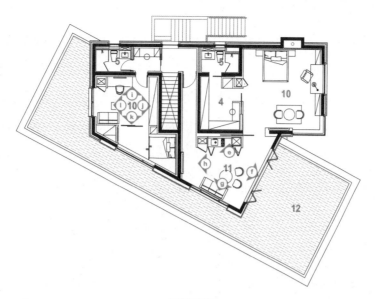

三层平面图

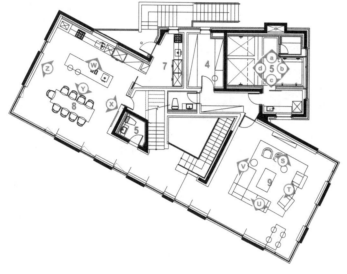

二层平面图

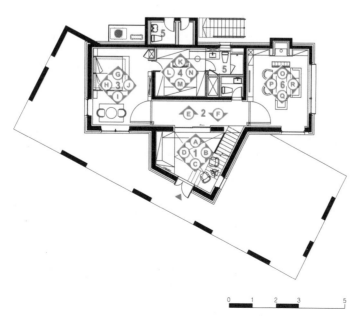

一层平面图

1. 入口
2. 走廊
3. 卧室
4. 衣帽间
5. 浴室
6. 会客室
7. 厨房
8. 餐厅
9. 起居室
10. 儿童房
11. 家庭娱乐室
12. 露台

It is a house that renovated the existing house of more than 10 years and was designed by the concept of 'The house to alive even after 100 years' as the demand of a client.

The building consists of a brick mass and exposed concrete mass, and each mass is divided into a private space and a public space. The two mass divides the space naturally and there is a staircase in the center of the overlapped building that the moving flow is designed as short as possible. Along the entrance of the second floor, there is a corridor connecting the bedroom and living room. The living room and bedroom are located on the two masses, and as the entire area is open to the open structure, the revolving door is placed for the use of a wall or a door. Along the diagonal staircase in the right side of the entrance, there are highest living room, kitchen, and bathroom on the second floor. Various sizes of windows along the wall enables to see the outside scenery from different angles, and the wall is finished with matt, and the floor is finished with polished marble, which adds the depth to the space. The bathroom located behind a hidden door in the living room was provided by the client's request, which enjoys sauna with close friends. This place points with white line patterns in dark marble with indirect lighting that creates a special atmosphere. The third floor with a family room with children's rooms brings a different atmosphere by using diagonal lines of ceiling from the existing architectural framework. A terrace is also for the space of family.

This place makes an unusual atmosphere that has no distinction between the inside and the outside by bringing the finishing materials from outside. Although the two masses are physically separated, they are connected to the inside that offers a new space.

这座别墅是翻新了10年前就建好了的别墅，在客户的要求下，以"可经历百年的别墅"为设计理念。

建筑包括砖结构部分和露石混凝土路部分，而每部分又分为私人间和公共空间。两部分将整体空间自然地分开，并且，在建筑重的中心位置设有楼梯，动线设计得尽可能短。顺着入口前行，可达二层，走廊将卧室及起居室连在一起。起居室与卧室位于两个同的部分，整个区域形成开放式结构，而旋转门则取代了墙体或通门。沿着倾斜的楼梯通过入口的右边，可以进入二层最高的起室、厨房和卧室。透过沿墙安装的不同规格的窗户可以从不同度欣赏外面的风景。通过使用毯子装饰墙面和抛光大理石装饰地增添了整体设计的深度。客户要求在门后设计一个隐藏的与朋友一起享受的桑拿浴室。在黑色大理石上点缀白色线条图案，并结间接照明的效果，营造出一种特别的氛围。三层的家庭房和儿童的天花板保留了原别墅的对角线性建筑结构。露台也是全家聚会好地方。

通过统一运用室内外的装饰材料，使这个地方没有明显的内外线，营造出一种不同寻常的氛围。虽然两部分在实体上是分开但是它们与内部相连接的地方又提供了一个新的空间。

外观

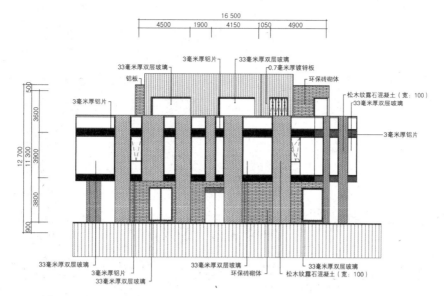

正立面图

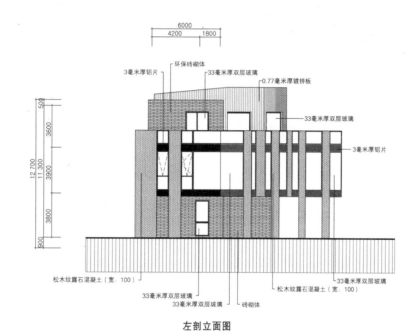

左剖立面图

右剖立面图

■ 入口

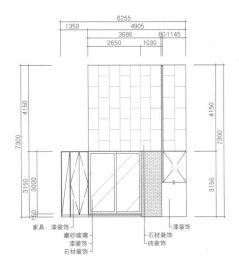

立面图A

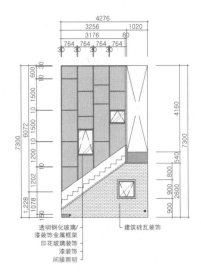

立面图B

立面图C

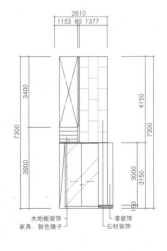

立面图D

走廊

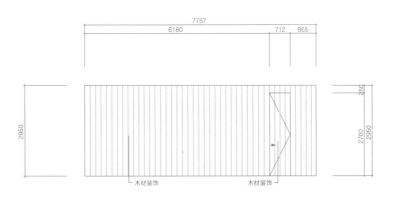

立面图E

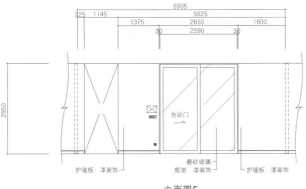

立面图F

■ 一楼卧室

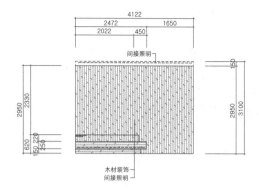

立面图G

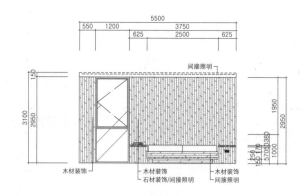

立面图H

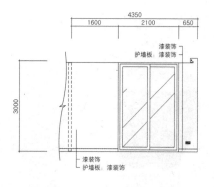

立面图I

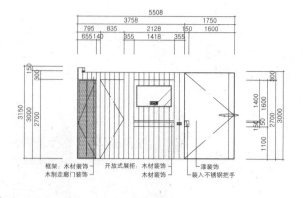

立面图J

衣帽间

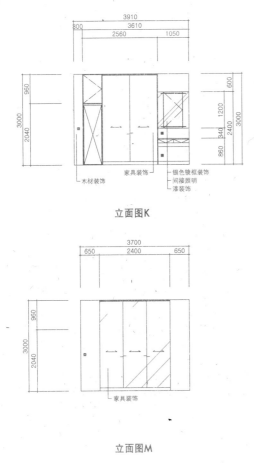

立面图K

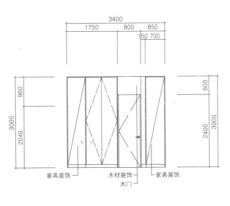

立面图L

立面图M

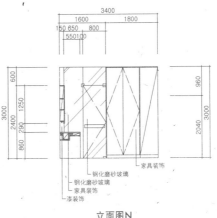

立面图N

■ 会客室

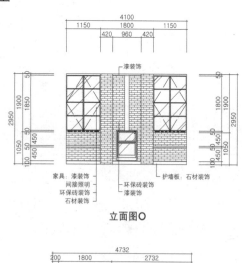

立面图O

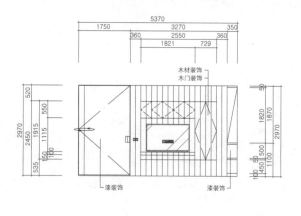

立面图P

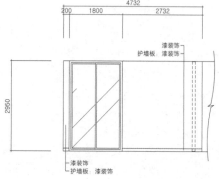

立面图Q

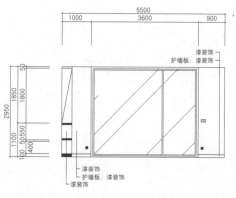

立面图R

二楼起居室

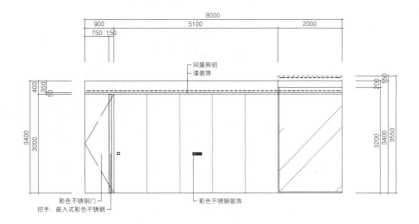

立面图S

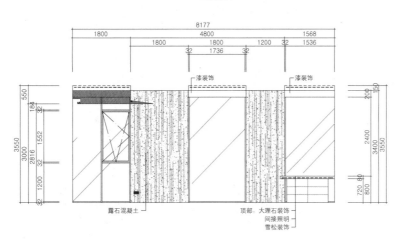

立面图T

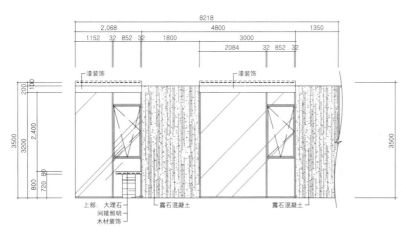

立面图U

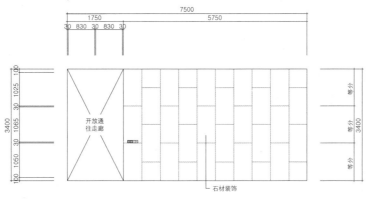

立面图V

厨房和餐厅

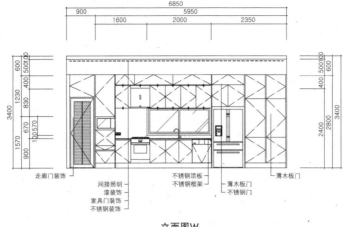

立面图W

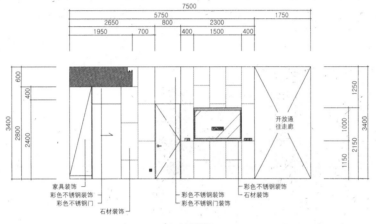

立面图X

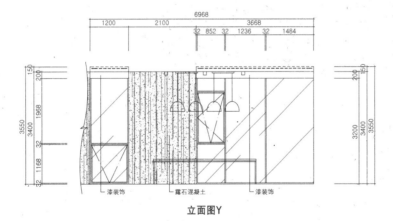

立面图Y

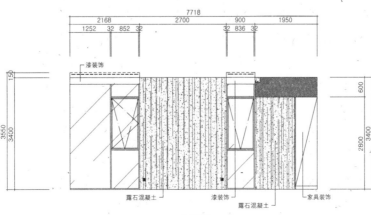

立面图Z

■ 二楼浴室

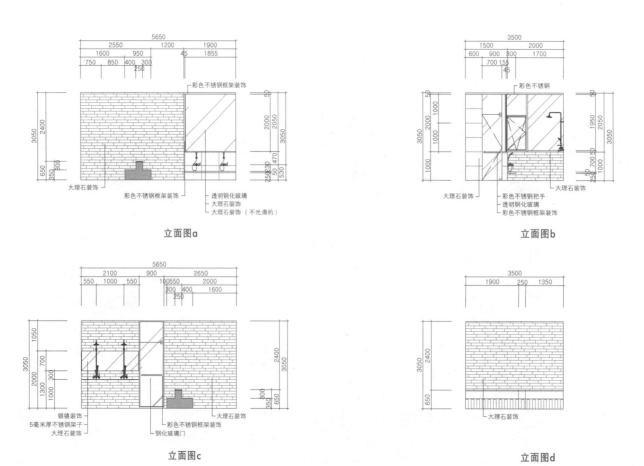

立面图a

立面图b

立面图c

立面图d

三楼家庭间

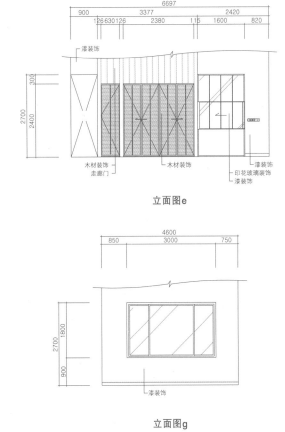

立面图 e

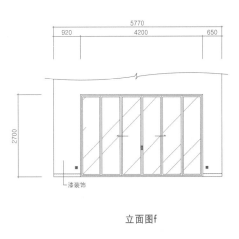

立面图 f

立面图 g

立面图 h

■ 三楼儿童房

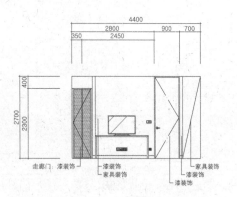

立面图i

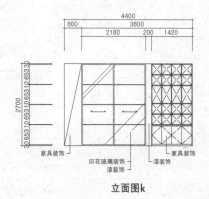

立面图k

立面图j

立面图l

楼梯

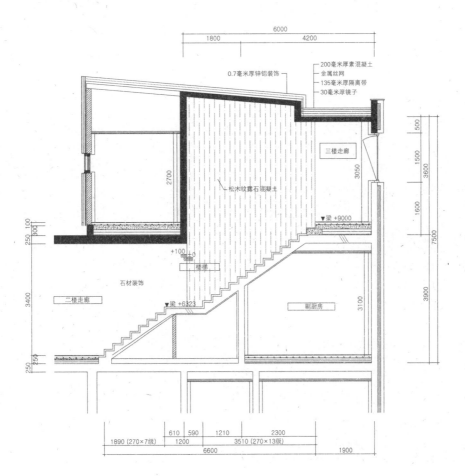

图书在版编目（CIP）数据

室内细部图集.2，商店与住宅 / 凤凰空间编．－－
南昌：江西科学技术出版社，2017.9
　ISBN 978-7-5390-5618-0

　Ⅰ．①室⋯　Ⅱ．①凤⋯　Ⅲ．①商店－室内装饰设计－
细部设计－图集②住宅－室内装饰设计－细部设计－图集
　Ⅳ．①TU238-64

中国版本图书馆CIP数据核字(2017)第153716号

国际互联网(Internet)地址：
http://www.jxkjcbs.com
图书代码：B17051-101
选题序号：KX2017089

责任编辑　　魏栋伟
特约编辑　　单　爽
项目策划　　凤凰空间 / 李文恒
售后热线　　022-87893668

室内细部图集2　商店与住宅　　　　　　　　　　　　凤凰空间　编

出版发行	江西科学技术出版社
社　　址	南昌市蓼洲街2号附1号　邮编：330009
	电话：(0791)86623491　86639342(传真)
印　　刷	北京博海升彩色印刷有限公司
经　　销	各地新华书店
开　　本	889 mm×1194 mm　1/16
字　　数	96千
印　　张	12
版　　次	2017年9月第1版　2023年3月第2次印刷
书　　号	ISBN 978-7-5390-5618-0
定　　价	198.00元

赣版权登字-03-2017-229
版权所有，侵权必究
（赣科版图书凡属印装错误，可向承印厂调换）